JANE'S POCKET BOOK OF RESEARCH AND EXPERIMENTAL AIRCRAFT

JANE'S POCKET BOOK OF RESEARCH AND EXPERIMENTAL AIRCRAFT

Compiled by Michael J. H. Taylor
Edited by John W. R. Taylor FRHistS, AFRAeS, FSLAET

Erratum:
The heading on page 35 should read **BELL XV-3** in place of **XB-3**

MACDONALD AND JANE'S

FIRST PUBLISHED 1976

COPYRIGHT © MICHAEL J. H. TAYLOR 1976

ISBN 0356 08409 4 (PVC EDITION)
ISBN 0356 08405 1 (LIBRARY EDITION)

This edition is not for sale in the United States, its dependencies, the Philippine Islands or the Dominion of Canada.

PUBLISHED BY MACDONALD AND JANE'S PUBLISHERS LIMITED
PAULTON HOUSE, 8 SHEPHERDESS WALK, LONDON N1 7LW

Printed in England by Netherwood Dalton & Co. Ltd., Huddersfield

FOREWORD

The 125 research and experimental aircraft described and illustrated in this Pocket Book are the most exciting of all the types flown in more than 30 years of aviation since the end of the Second World War. They reflect a determination that the tremendous impetus in the development of aircraft that had come about through strife should not be allowed to dry up (as had happened in 1919) once the immediate necessity for weapons to feed a fighting force had gone. Furthermore, the way in which America had forced Japan to surrender, and the new division of countries, had made it clear that no major force could ever again sit back on its laurels and watch the world go by.

It would, however, be a gross mistake to believe that all post-war research and experiments were carried out for military ends. Although many of them were, a great deal of time, energy and resource was and is channelled to other ends, including the development of new-technology airliners, greater safety in the air, new and more efficient wing shapes and space shuttle research.

Some of the aircraft covered in this book had origins that can be traced back to Nazi Germany — technology that could not be ignored by post-war powers. Others were out-and-out world-beaters, like the X-1 in which "Chuck" Yeager became the first man to fly faster than the speed of sound, and the X-15 that recorded an incredible 4,534 mph twenty years later and that reached an altitude of 354,200 ft. More unusual aircraft include the Northrop flying wings, one of which was built for ramming the tails of enemy aircraft, parasite fighters and a multitude of others. European aircraft include the French ramjet-powered double-delta Griffon and the British BAC 221 and HP 115 narrow deltas which helped to prove the practicability of the Concorde design. An interesting inclusion in this book is the TSR.2. Although built as a prototype combat type, the TSR.2 never entered production and, because of the use of very advanced technology, can be permitted to be listed among experimental designs.

One of the main preoccupations of the three decades covered in this book was the pursuit of a successful VTOL (Vertical Take-Off and Landing) aircraft and here, too, is the whole range of tilt-wing, tilt-rotor, tilt-duct, tail-sitting, fan-in-the-wing, deflected-thrust and other research concepts that have pioneered VTOL. From them have evolved the incredible Harrier combat aircraft, and a new generation of high-speed commercial and military helicopters. Among the most important types in this Pocket Book are the lifting-body research aircraft on which America has been engaged since the 1950s. These strange bulbous little wingless aircraft have produced data for the Space Shuttle craft, and it was with one of these, the Martin Marietta X-24B, that the last known powered flight by a rocket-powered aircraft was made in September 1975.

Never before have these advanced and historic aircraft been covered in a single volume. The name Jane's is sufficient guarantee of the accuracy of the text. The full-page illustrations are the finest available, gathered from every part of the aircraft-manufacturing world. Not least, this Pocket Book includes Soviet research aircraft like the Mikoyan Analogue, which flight tested a small-scale version of the wing of Russia's counterpart to the Concorde before the full-size aircraft left the ground.

M. J. H. TAYLOR

(France)

AÉROCENTRE N.C. 3020 BELPHÉGOR

First flight: 6 June 1946
Purpose: Experimental monoplane for stratospheric research
Power plant: One Daimler-Benz DB610 engine, consisting of two DB605 twelve-cylinder engines coupled together (max output 3,000 hp)
Wing span: 73 ft 3 in (22.32 m)
Length: 58 ft 8¾ in (17.90 m)
Wing area: 538.2 sq ft (50 m²)
Weight loaded: 22,046 lb (10,000 kg)
Max level speed at 19,675 ft (6,000 m): 297 knots (342 mph; 550 km/h)
Service ceiling: 42,000 ft (12,800 m)
Take-off run: 1,410 ft (430 m)
Accommodation: Pressurized accommodation for crew of five, including two research crew members. Pressure cabin consisted of cylinder 5 ft 7 in in diameter x 17 ft 4¾ in long (1.7 m x 5.3 m) installed in centre-fuselage. Pilot had raised cupola above cabin. Observation windows in fuselage under wing trailing-edge.
Special design features: Cantilever wings of composite structure, consisting of two half centre-sections and two outer wings. Chord (at 11 ft 5½ in; 3.5 m from fuselage centreline) 9 ft 2¼ in (2.8 m); thickness/chord ratio 16%. Composite-structure fuselage in three sections; nose-section carrying engine; all-metal centre-fuselage; and wooden rear section. Single fin and rudder. Trim-tabs in rudder, elevators and ailerons. Retractable tailwheel-type landing gear. Annular radiator in front of engine. Four-blade propeller of 14 ft 9¼ in (4.5 m) diameter.
History: The N.C. 3020 was a development of the Farman 1000 of 1932 and was designed specially for stratospheric research. It was built at the Billancourt works and taken to Toussus-le-Noble for assembly, after which the research programme was initiated.

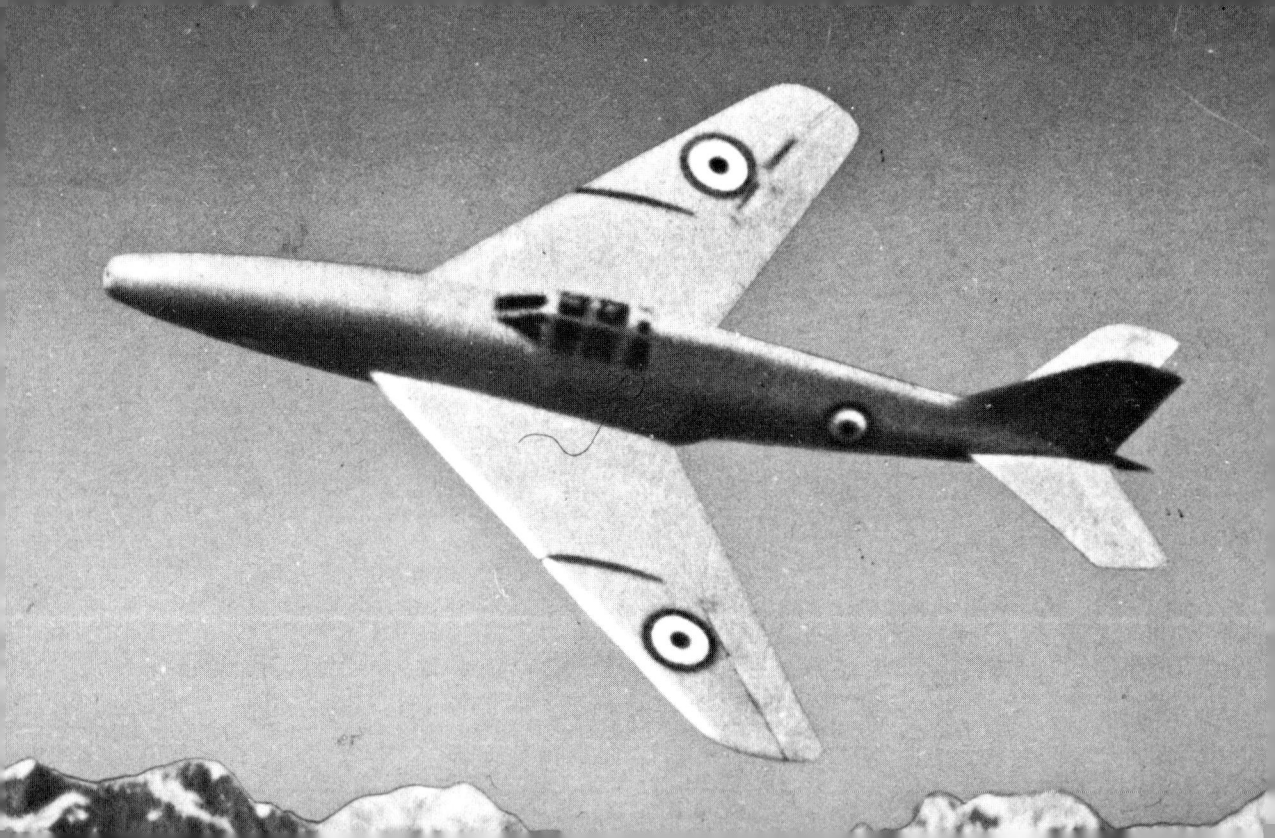

(Italy)

AMBROSINI SAGITTARIO

First flight: 5 January 1953
Purpose: Experimental aircraft for aerodynamic research
Data: First aircraft
Power plant: One Turboméca Marboré II turbojet engine (840 lb; 380 kg st)
Wing span: 24 ft 7¼ in (7.50 m)
Length: 30 ft 7 in (9.32 m)
Wing area: 157.15 sq ft (14.6 m²)
Max level speed at 13,125 ft (4,000 m): 302 knots (348 mph; 560 km/h)
Service ceiling: 26,250 ft (8,000 m)
Range: 307 nm (354 miles; 570 km)
Accommodation: Pilot only
Special design features: Wooden construction. Basic fuselage of the Ambrosini Super 7 piston-engined advanced trainer. Engine mounted in nose, with direct air entry and with jet exit beneath fuselage. 45° sweepback on wings, and swept tail surfaces. Tailwheel landing gear with inward-retracting main wheels.
History: Following initial flight trials with the Freccia (the Sagittario airframe powered by an Alfa Romeo piston engine), the first Sagittario made its maiden flight as a jet-powered aircraft of wooden construction. The use of wood was regarded as an improvisation, to obtain flight experience with the sweptwing configuration with as little delay as possible. It was considered that the aircraft, after the necessary modifications and installation of a new power plant, would fly faster than the speed of sound. An all-metal single-seat lightweight fighter version, known as the Sagittario II and powered by a Rolls-Royce Derwent turbojet engine, was built by Aerfer and proved capable of diving at Mach 1.1.

(UK)

ARMSTRONG WHITWORTH A.W.52

First flight: 13 November 1947
Purpose: Research into tail-less aircraft, and to obtain data to determine the final shape and detailed design of a six-jet tail-less transport aircraft
Data: First prototype
Power plant: Two Rolls-Royce Nene turbojet engines (each 5,000 lb; 2,270 kg st)
Wing span: 90 ft 0 in (27.43 m)
Length: 37 ft 4 in (11.38 m)
Height: 14 ft 5 in (4.39 m)
Wing area: 1,314 sq ft (122.07 m²)
Weight loaded: 34,154 lb (15,492 kg)
Max level speed: 434 knots (500 mph; 805 km/h)
Service ceiling: 50,000 ft (15,240 m)
Cruising range at 36,000 ft (10,975 m): 1,302 nm (1,500 miles; 2,414 km)
Accommodation: Crew of two
Special design features: Wing in three main sections: a centre section with sweptback leading-edge and two tapered and sweptback outer sections. Sweepback of 35° on the leading-edge and 24° 45' at quarter-chord. Longitudinal and lateral control by "controllers" (elevons), hinged on each outer wing to serve as both elevators and ailerons. Controllers hinged to movable surfaces known as "correctors", which provided fore-and-aft trim and were used to counteract pitching caused by the lowering and raising of the flaps. Twin elliptical fins and rudders for directional control mounted on outer wings. Crew compartment projected from leading-edge of centre-section, with accommodation for pilot (with ejector seat) and navigator/radio operator. Anti-spin chutes carried in wingtip containers.
History: Following on from research carried out with the A.W.52G glider, two prototype powered aircraft were built to Specification E.9/44. The first to fly (at Boscombe Down) was powered by two Nene engines; it was followed on 1 September 1948 by the second aircraft, with Rolls-Royce Derwent 5 engines (each 3,500 lb; 1,588 kg st). The first aircraft was involved in an accident caused by asymmetric flutter on 30 May 1949 and, although not badly damaged, was abandoned. The other A.W.52 completed a research programme with the RAE at Farnborough.

(France)

ARSENAL VG 70

First flight: 23 June 1948
Purpose: Research for high-speed jet fighters
Power plant: One Junkers Jumo 004B-2 turbojet engine (1,890 lb; 857 kg st)
Wing span: 29 ft 10¼ in (9.10 m)
Length: 31 ft 9¾ in (9.70 m)
Height: 7 ft 6½ in (2.30 m)
Wing area: 182.9 sq ft (17.0 m²)
Weight loaded: 7,496 lb (3,400 kg)
Max level speed at 22,965 ft (estimated): 485 knots (559 mph; 900 km/h)
Accommodation: Pilot only
Special design features: All-wooden wings with rounded tips, swept-back at 38° at quarter-chord. Wing dihedral of 6°. Perforated dive-brakes on upper surfaces and built-in leading-edge pillar-box slots. Slotted trailing-edge flaps between short-span ailerons and fuselage. Air intake under fuselage.
History: Designed by M. Galtier, the VG 70 was built by the Arsenal de l'Aéronautique, which had been formed in 1936 under the law for the nationalisation of military industries. By 1947, the aircraft had been completed, but flying was delayed until June 1948 because of aerodynamic troubles shown up during wind-tunnel testing. When flying started the aircraft proved to have a comparatively high speed, considering the low power of its engine, thanks to its small proportions. As noted in the *Jane's All the World's Aircraft* of the time, it was intended that the axial-flow Jumo engine should be replaced eventually with a Rolls-Royce Derwent 5. This change of power plant would have required redesign of the fuselage, to accept the increased width of the centrifugal unit, and the idea was abandoned.

(UK) AVRO ASHTON

First flight: 1 September 1950
Purpose: Flying test-bed, designed to undertake research in some phases of high-altitude jet operation
Power plant: Four Rolls-Royce Nene 5 and 6 turbojet engines (each 5,000 lb; 2,268 kg st)
Wing span: 120 ft 0 in (36.58 m)
Length: 89 ft 6½ in (27.29 m)
Height: 31 ft 3 in (9.53 m)
Wing area: 1,421 sq ft (132 m²)
Weight loaded: up to 82,000 lb (37,195 kg)
Max level speed at 20,000 ft (6,100 m), at 72,000 lb (32,659 kg) AUW: 380 knots (438 mph; 705 km/h)
Service ceiling (at above weight): 40,500 ft (12,345 m)
Still air range (at above weight): 1,498 nm (1,725 miles; 2,775 km)
Accommodation: Crew of five

Special design features: Adapted from the Avro Tudor Mk 2 airframe, with a tricycle landing gear, revised tail unit and four Rolls-Royce turbojet engines. Engines paired in large nacelles, one beneath each wing.
History: A successor to the Tudor 8, the Ashton was known originally as the Tudor 9. Six were built for the Ministry of Supply, and were unusual in being large aircraft intended solely for research. The first Ashton, designated Mk I, was used to collect information on turbojet flying at high altitude. The second aircraft was operated for air-conditioning tests; the third differed in that it carried a radome and underwing containers for bombs, and was used to develop advanced bomb-sight equipment. The remaining aircraft were made up of two additional Mk IIIs and a single Mk IV, and were used for developmental engine and instrument tests, visual bombing and ballistics research, and de-icing trials. One was used as the test-bed for the Bristol Olympus afterburning turbojet engine.

(UK) AVRO TYPE 707

First flight: 4 September 1949
Purpose: Low- and high-speed delta-wing research
Photo: Type 707C
Power plant: One Rolls-Royce Derwent turbojet engine (3,600 lb; 1,633 kg st)
Wing span: 707A and C 34 ft 2 in (10.41m)
707B 33 ft 0 in (10.06m)
Length: 42 ft 4 in (12.90 m)
Height: 707A 11 ft 7 in (3.53m)
707B 11 ft 9 in (3.58m)
Weight loaded: 707C 10,000 lb (4,536 kg)
Accommodation: Pilot only in all versions except 707C, which had side-by-side seating for a crew of two, with dual controls.
Special design features: One-third scale model of the Type 698 Vulcan bomber. Delta wings. Dorsal air intakes and low-speed controls on types 707 and 707B. Wing-root air intakes and modified control surfaces on 707A. Wider cockpit canopy on 707C.

History: Avro Type 707s were built primarily to carry out extensive research into the behaviour of delta wings at low speeds and to provide information which would be used in the final design of the Vulcan, the first jet bomber to have a delta wing. First British aircraft designed for delta-wing research, the Type 707 made its maiden flight at Boscombe Down, and was destroyed subsequently in an accident. It was followed by the Type 707B, also built for low-speed research, which first flew on 6 September 1950. The third and fourth aircraft were Type 707As, built for high-speed research, and these made their maiden flights on 14 June 1951 and on 20 February 1953. Final aircraft of the series was the Type 707C, a two-seat dual-control version of the Type 707A constructed to give pilots experience in flying delta-winged aircraft. The first flight of the Type 707C was made on 1 July 1953, and it was still in the RAE, Farnborough, ten years later.

(UK)

BAC 221

First flight: 1 May 1964
Purpose: Research aircraft
Power plant: One Rolls-Royce Avon RA.28R turbojet engine with afterburning (about 14,000 lb; 6,350 kg st)
Wing span: 25 ft 0 in (7.62 m)
Length: 57 ft 7½ in (17.56 m)
Height: 13 ft 0 in (3.96 m)
Wing area: 504 sq ft (46.82 m²)
Max T-O weight: 18,000 lb (8,165 kg)
Max level speed: 921 knots (1,060 mph; 1,706 km/h)
Service ceiling: 50,000 ft (15.240 m)
Accommodation: Pilot only
Special design features: Slender ogival delta wings. Aspect ratio 1.22. Thickness/chord ratio 4.5%. Sweepback on leading-edge over main portion of each wing 65°, curving round to streamwise tip. Entire trailing-edge of wings made up of ailerons and elevators. Fuselage fitted with hydraulically-actuated drooping nose. Four petal airbrakes around jet nozzle. Sweptback fin and rudder.
History: Under Ministry of Aviation contract, the Filton Division of BAC redesigned and almost completely rebuilt the Fairey Delta 2 research aircraft which had once held the World's air speed record. In its revised form, as the BAC 221, the aircraft had new delta wings, control surfaces, engine intakes and landing-gear, and was instrumented and equipped for use by the Royal Aircraft Establishment, Bedford, for a basic research programme. The fuselage was also lengthened. Following the initial development flying, a further modification programme was put in hand. This involved the fitting of a power-jettisonable clear-view canopy, an extended fin to improve stability, anti-spin parachute for use during slow-speed research flying, a new compensated pitot static probe for instrumentation measurements, and other equipment. The rebuilt aircraft was delivered to the RAE on 20 May 1966. There it was used for research over the subsonic, transonic and supersonic speed ranges up to about Mach 1.6. It also extended the scope of investigations with slim delta wings at low speeds. To study airflow patterns and vortex formations, two research methods were used: pressure plotting and "tufting". Also, because of the basic similarity between the wing planforms of the 221 and the Concorde airliner, the research had an important bearing on the latter's development programme. The BAC 221 was last flown on 9 June 1973, and was acquired subsequently by the Museum of Flight at East Fortune, near Edinburgh. During its active life the aircraft made 273 flights totalling 139¾ flying hours.

(UK)

BAC TSR 2

First flight (XR219): 27 September 1964
Purpose: Attack and reconnaissance aircraft
Power plant: Two Bristol Siddeley Olympus 22R turbojet engines (each 30,610 lb; 13,885 kg st with afterburning)
Wing span: 37 ft 0 in (11.28 m)
Length overall: 89 ft 0 in (27.13 m)
Height: 24 ft 0 in (7.32 m)
Normal max T-O weight (estimated): 95,900 lb (43,500 kg)
Max level speed (estimated): Mach 2.05 — 2.5
Max rate of climb at S/L: over 50,000 ft (15,240 m)/min
Range, with underwing tanks (estimated): 1,500 nm (1,727 miles; 2,780 km)
Accommodation: Crew of two
Special design features: High-mounted wings with turned-down wingtips. Wings of very low thickness/chord ratio, sweptback at 60° on leading-edges. Aerofoil varied, particularly at root, to ensure optimum efficiency. Full-span blown flaps, except on wingtips. All control surfaces at tail. One-piece all-moving vertical tail surface. All-moving horizontal surfaces ("tailerons") operated together for pitch control and differentially for roll control. Each horizontal surface had a separate control surface inset in its trailing-edge, but these were locked in high-speed flight. Engine air intakes of variable-area type with movable half-cone shock-bodies.
History: The specification of the TSR 2 was based originally on Operational Requirement 339 for an aircraft to replace the Canberra in its long-range interdiction and reconnaissance roles. The TSR 2 far exceeded these requirements. It proved suitable for all types of attack, with weapons ranging from rocket projectiles to high-yield nuclear missiles, and might also have been used, had it not been cancelled, in the strategic deterrent role. XR219, the only example to fly, was the first of 20 projected development and pre-production models. A contract for an additional 30 TSR 2s had also been negotiated. The navigation-attack system of the TSR 2 was the most advanced fitted, at that time, to any aircraft developed for Western air forces. Completely automatic sorties, including attack at high or low level, would have been possible, without any visual reference; although the crew would monitor all phases of the mission and the pilot could take over control at any time. All systems were fail-safe and the aircraft would enter a climb automatically in the event of failure of the terrain-following system. The accuracy of its weapon delivery capability was officially stated to be within "tens of feet". The last flight of the prototype was made on 31 March 1965, 24 flights having been made totalling 13 hours 14 minutes flying time.

(UK)

BAC (HUNTING) H.126

First flight: 26 March 1963
Purpose: Jet-flap research aircraft
Power plant: One Bristol Siddeley Orpheus turbojet engine
Wing span: 45 ft 4 in (13.82 m)
Length over nose-probe: 50 ft 2 in (15.29 m)
Height: 15 ft 6 in (4.72 m)
Accommodation: Pilot only
Special design features: Wings with two alternative dihedral angles of 4° and 8°. Ailerons and flaps had slots in leading- and trailing-edges, through which cooling air was passed. Narrow slit along trailing-edge of wings for ejection of engine efflux. Tailplane mounted near top of fin. Tailplane incidence variable hydraulically, in conjunction with elevator. Trim and anti-balance tab in rudder. Anti-spin parachute in top of fin. Non-retractable nosewheel-type undercarriage (see below for other features).
History and operation: Announced officially in February 1961, the H.126 was a research aircraft built to flight-test the jet-flap principle. In this, the greater part of the efflux from the aircraft's turbojet engine was ducted into the wings and ejected through a narrow slit along the trailing-edge, as a thin gaseous sheet over the flaps, which acted as jet-stream deflectors. The stream followed the flaps as they were lowered, forming a "jet-flap" of high-velocity gas, which offered theoretical lift coefficients of 10 or more. This part of the efflux produced thrust as well as lift, and further thrust was provided through low-set nozzles about mid-way along each side of the fuselage. Jet nozzles in the tail controlled pitch and yaw. Nozzles in the wingtips, to control roll, were operated by an autostabiliser. The elevators were so linked to the tailplane that when tailplane incidence increased the elevators moved down (and vice-versa).

The H.126 was fitted with extensive test instrumentation, most of the rear fuselage being occupied by automatic observer equipment, etc. Over 100 test flights were completed by mid-1965.

(USA)

BELL X-1 series

First flight (powered): 9 December 1946
Purpose: Investigation of supersonic flight problems
Photo: X-1A
Data: X-1
Power plant: One Reaction Motors E6000-C4 (Thiokol XLR-11) bi-propellant rocket motor (6,000 lb; 2,722 kg st)
Wing span: 28 ft 0 in (8.53 m)
Length: 31 ft 0 in (9.45 m)
Height: 10 ft 10 in (3.30 m)
Weight loaded: 13,400 lb (6,078 kg)
Max level speed: 831 knots (957 mph; 1,540 km/h)
Accommodation: Pilot only
Special design features: Rocket engine and thin-section straight wings. X-1B intended primarily for research on thermal problems encountered in high-speed flight, and specially equipped and instrumented for the purpose.
History: Three X-1s were built, the type first being air-launched unpowered, from a Boeing B-29 Superfortress on 19 January 1946. Powered flights began in December of the same year and on 14 October 1947 the first X-1, piloted by Captain Charles Yeager, became the first aeroplane to exceed the speed of sound, reaching 582 knots (670 mph; 1,078 km/h, or Mach 1.015) at an altitude of 42,000 ft (12,800 m). The second X-1 was used by the NACA for high-speed flight research; the third aircraft was destroyed at Edwards AFB during fueling operations. The X-1A was similar to the X-1, except for having turbo-driven fuel pumps (instead of a system using nitrogen under pressure), a new cockpit canopy, longer fuselage and increased fuel capacity. In this aircraft a speed of Mach 2.435 was achieved on 12 December 1953; in the following June an altitude of over 90,000 ft was reached. In 1955, this aircraft was given new wing panels, but was destroyed before its first flight in this configuration. Following the X-1B, used for thermal research, came the projected X-1C, which was cancelled, and the X-1D. The latter aircraft was destroyed in August 1951 after being jettisoned from its B-50 carrier-plane, following an explosion. The last of the series was the X-1E. This was the second of the original X-1s fitted with wings of 4% thickness/chord ratio (instead of 10%), turbo-driven fuel pumps and a knife-edge windscreen. Ballistic control rockets, designed by Bell Aircraft's Rocket Division, were included in this aircraft, which was flight tested by the NACA. This system ensured adequate control at extreme altitudes. A total of 156 flights were made with the X-1, 21 with the X-1A, 27 with the X-1B, one with the X-1D and 26 with the X-1E.

(USA) BELL X-2

First flight (second aircraft, powered): 18 November 1955
Purpose: To explore the problems of transonic and supersonic flight
Power plant: One Curtiss-Wright XLR25-CW-1 throttlable liquid-propellant rocket motor (15,000 lb; 6,804 kg st)
Wing span: 32 ft 0 in (9.75 m)
Length: 44 ft 0 in (13.41 m)
Height: 13 ft 6 in (4.11 m)
Max level speed: Mach 3.2
Accommodation: Pilot only
Special design features: K-monel metal cylindrical fuselage. Stainless-steel swept wings and tail. Single nosewheel and underfuselage flat skid landing gear; underwing balancer skids able to be fitted for low-speed flights. Pilot's cockpit detachable in an emergency, being carried, after separation from the rest of the aircraft, to a lower altitude by a ribbon parachute, at which height the pilot could bale out.
History: Two prototypes of the X-2 were developed jointly by Bell, the USAF and the NACA. The first was lost in May 1954 when it was jettisoned from its B-50 carrier aircraft after being damaged by an explosion in the B-50. The second X-2 made its first powered flight piloted by Lt-Col F Everest of the USAF. Following seven other flights, it was taken over by Capt Iven Kincheloe, who, on 7 September 1956, attained a height of 126,200 ft (36,637 m). This aircraft was destroyed in a fatal crash on 27 September 1956, after a flight in which its pilot, Capt Milburn Apt, USAF, had flown at a hitherto-unprecedented Mach 3.2 (1,818 knots; 2,094 mph; 3,370 km/h).

(USA) BELL X-5

First Flight: 20 June 1951
Purpose: Investigate the aerodynamic effects of changing the degree of wing sweepback during flight
Power plant: One Allison J35-A-17 turbojet engine (4,900 lb; 2,222 kg st)
Wing span (unswept): 30 ft 9¾ in (9.39 m)
Wing span (swept): 18 ft 7 in (5.66 m)
Length: 33 ft 4 in (10.16 m)
Height: 12 ft 0 in (3.66 m)
Weight loaded: 9,892 lb (4,487 kg)
Max level speed: 525-564 knots (605-650 mph; 973-1,046 km/h)
Accommodation: Pilot only
Special design features: Variable-sweepback wings, the operating mechanism compensating for the resultant shift of the centre of gravity. Each wing root had a specially designed fairing to ensure that the root leading-edge and trailing-edge presented a smooth aerofoil surface regardless of the angle of sweepback. Sweepback angle of the wings could vary from 20° to 60° maximum. Full-span leading-edge slats. Hydraulically-operated dive brakes located in fuselage sides forward of cockpit.
History: Design work on the Bell X-5 started in 1948, following the decision to construct an aircraft based on the Messerschmitt P.1101 of wartime concept. It was appreciated that by having such a wing arrangement the otherwise-poor handling qualities of heavily swept aircraft at low speeds could be improved by reducing the angle of sweep and so giving them the handling characteristics of aircraft with straighter wings, while still retaining the performance that a heavily swept wing can offer. Both X-5s were later flown at Edwards Air Force Base, the first being operated by the US Air Force until it crashed on 13 October 1953, killing its pilot. The second aircraft was flown by the NACA.

(USA) BELL X-14

First flight (hovering flight): 19 February 1957
Purpose: Experimental vertical take-off aircraft
Power plant: Two Bristol Siddeley Viper turbojet engines
Wing span: 34 ft 0 in (10.36 m)
Length: 25 ft 0 in (7.62 m)
Height: 8 ft 0 in (2.44 m)
Max level speed: about 139 knots (160 mph; 257 km/h)
Accommodation: Pilot only
Special design features: Simply-designed aircraft with open cockpit, straight wings and two engines mounted side-by-side in fuselage nose. Thrust diverters mounted behind engines to deflect the jet efflux towards the ground during take-off and landing, or rearward for forward thrust. Compressed-air nozzles at wingtips and tail for low-speed or hovering directional and stability control.
History: The X-14 was an experimental VTOL research aircraft of the jet-deflection type, which was built under USAF contract. All flight development up to the Summer of 1960 was done with the aircraft in its original form, powered by Viper engines. These were replaced by two General Electric J85 turbojets under a NASA programme to evaluate the J85 as a power plant for VTOL applications.
During take-off and landing, the thrust diverters deflected the jet efflux towards the ground, thus enabling the X-14 to raise itself vertically off the ground by direct jet lift, with its fuselage horizontal. At a safe height, the jet efflux was directed slightly rearward to provide some forward thrust in addition to jet lift. When the forward speed was sufficient for the fixed wings to provide adequate lift, the efflux was redirected towards the rear and the aircraft flew in a conventional manner. The X-14 made its first hovering flight on 19 February 1957 and first successful transition from hovering to forward flight on 24 May 1958.

(USA)

BELL X-22A (MODEL D2127)

First flight: 17 March 1966
Purpose: Tilting-duct V/STOL research aircraft
Power plant: Four General Electric YT58-GE-8D shaft-turbine engines (each 1,250 shp), mounted in pairs at root of each wing
Wing span: 39 ft 3 in (11.96 m)
Span across front ducts: 23 ft 0 in (7.01 m)
Length (excl boom): 39 ft 7 in (12.07 m)
Height: 20 ft 8 in (6.31 m)
VTOL max weight: 16,274 lb (7,381 kg)
Max level speed at S/L (estimated): 275 knots (316 mph; 510 km/h)
Range (estimated): 386 nm (445 miles; 715 km)
Accommodation: Crew of two, plus passengers, cargo or mixed loads
Special design features: Wings mounted at rear of fuselage, each with a tilting duct built into leading-edge. Elevon control surface in each wing. Square-section fuselage. Fixed vertical fin at tail, with no rudder. Foreplane at front of fuselage, carrying two forward ducts. Elevon control surface in slipstream from each duct, as on wings. Cross-shafting of engines ensured that all ducted propellers continued to be powered in the event of an engine failure. Control of aircraft achieved at ducts through thrust modulation, obtained by propeller pitch change, and by means of the four elevons, in the slipstream of each duct. Altitude control in vertical flight by increasing and decreasing engine power.
History: Under a US Navy contract, awarded in November 1962, Bell Aerospace Company built and flew two prototypes of this tri-service research aircraft. The first flew in March 1966, making four vertical flights to a height of 25 ft (7.6 m) and turning through 180°. Subsequently, it made several STOL take-offs with the ducts at an angle of 30°, followed by forward flights at 100 mph, before suffering severe damage in a heavy landing on 8 August 1966. It was considered beyond repair. The second X-22A flew on 26 January 1967, and by May 1969 had accumulated 110 hours flying time, during which it made 386 vertical and 216 short take-offs, 405 vertical and 197 short landings, and 185 transitions from vertical to horizontal flight and vice versa. In the Spring of 1968, a variable stability system (VSS), designed and built by Cornell Aeronautical Laboratory, was installed. This enabled the aircraft to change its flight characteristics automatically while airborne, making possible further research. On 19 May 1969, the X-22A was formally delivered to the US government and on 18 January 1971 it was transferred to the Cornell Aeronautical Laboratory where further tests of the VSS were carried out.

(USA)

XB-3

First flight (vertical): 23 August 1955
Purpose: Experimental tilting-rotor convertiplane
Power plant: One Pratt and Whitney R-985 piston engine (450 hp)
Rotor Diameter: 33 ft 0 in (10.06 m)
Span of fixed wings: 31 ft 3½ in (9.54 m)
Length of fuselage: 30 ft 3½ in (9.23 m)
Height: 13 ft 6 in (4.11 m)
Effective wing area: 120 sq ft (11.15 m²)
Normal T-O weight: 4,800 lb (2,177 kg)
Max level speed: about 157 knots (181 mph; 291 km/h)
Accommodation: Seating for four

Special design features and History: The XV-3 was developed by Bell and the USAF Research and Development Command, for the US Army, under a joint Army-Air Force contract initiated in 1951.

Two combination rotor/propellers, mounted near the tips of the fixed wings, operated as conventional lifting rotors during take-off, landing and low-speed flight; they tilted to a forward-facing position for horizontal cruise and high-speed flight, once lift had been transferred to the wings. The rotors were tilted by electric motors enclosed in the fairings at each wingtip; movement through the full range from horizontal to vertical attitudes could be accomplished in 10 to 15 seconds. Originally, three-blade articulated rotors were fitted; but a two-blade semi-rigid rotor system, with underslung feathering axis hubs, was used after 1957. The XV-3 made its first vertical flight in August 1955 and had made 15° conversions in flight before it was badly damaged in a crash landing on 25 October 1956. Development continued with a second XV-3, which was subjected to full-scale wind-tunnel tests before entering its flight test programme. The first full conversion from vertical to horizontal flight, and vice versa, was made on 18 December 1958. More than 110 full conversions followed, from all flight attitudes, and the XV-3 was evaluated by NASA. An XV-3 modification contract was awarded to Bell by NASA in February 1962, for design, development and evaluation of engineering changes to the rotor system to improve high-speed stability and control of the convertiplane. It remained in a hanger until 1965 while engineers worked on a solution. During May 1966, however, a total of 25 test runs were conducted successfully. The aircraft was subsequently damaged due to failure of a pylon mounting. The XV-3 was the world's first tilting-rotor, fixed-wing aircraft to achieve 100% tilting of its rotors and proved the design practical. In all the aircraft completed more than 250 flights, totalling more than 125 flying hours.

(UK)

BOULTON PAUL P.111 and P.120

First flight: 10 October 1950 and 6 August 1952 respectively
Purpose: To investigate the delta wing at transonic speeds
Photo and Data: P.111
Power plant: One Rolls-Royce Nene turbojet engine (5 100 lb; 2 313 kg st)
Wing span: 25 ft 0 in — 33 ft 5½ in (7.62 m — 10.20 m)
Length: 26 ft 1 in (7.95 m)
Height: 12 ft 6½ in (3.82 m)
Wing area: 200 sq ft (18.58 m^2)
Weight loaded: about 9 600 lb (4 354 kg)
Max level speed: Mach 0.95-0.98
Accommodation: Pilot only
Special design features: Delta wings, swept at 45° on the leading-edges, with very low thickness/chord ratio. Detachable wingtips to enable tests to be carried out with blunt or more-pointed tips. Large-area pointed vertical tail with detachable tip. Broad air intake in nose. Wide-track landing gear, witm main wheels retracting inwards. Parachute to reduce landing speed and run, carried in fairing on port side of rear fuselage. P.111A version had four fuselage-mounted airbrakes and nose probe containing a pressure head.

Designed to specification E27/46, the P.111 made its first flight at Boscombe Down in October 1950. During a period of tests, the aircraft was landed with the widercarriage retracted owing to failure of the undercarriage door operating mechanism. While repairs were being made the P.111 airframe was modified internally and externally, and it was redesignated P.111A. In this configuration the aircraft first flew on 2 July 1953 and thereafter continued its test programme. Boulton Paul were responsible for flight testing of the P.111A until 14 January 1954, during which time 40 flights were made. It was then delivered to the R.A.E. Meanwhile, the P.120 had been built and flown for the first time on 6 August 1952. This differed from the P.111 in having an all-moving tailplane, with a fin and rudder of swept and near-parallelogram configuration to accommodate the new tail. It also featured new dive brakes, later to be used on the modified P.111A, and a new fairing between the vertical tail and the jet orifice for the parachute. The P.120 was destroyed in an accident caused by tail flutter during a test flight on 28 August 1952. The P.111A made its last flight with the Aero Flight of the R.A.E. on 20 June 1958, after which it was passed to the College of Aeronautics, Cranfield for apprentice training

(USSR)

BRATUKHIN OMEGA

First flight: Tested 1941
Purpose: Twin-engined twin-rotor research helicopter
Photo: Bratukhin G-4, basically similar to the Omega except for its engines
Power plant: Two MV-6 radial engines (each 220 hp)
Forward speed: 97 knots (112 mph; 180 km/h)
Vertical climb: 1,200 ft/min (360m/min)
Accommodation: Crew of two in tandem
Special design features: Metal tubular fuselage structure, fabric covered, with high-mounted cockpit affording 360° vision for crew. Two three-blade inward-rotating rotors, driven by vertically-mounted shaft extensions from engines. Power units and rotor masts supported by open tubular Warren girder structure mounted on top of fuselage. Cantilever tailfin and rudder, with tailplane set on top of fin. Landing gear wheels under nose and tail of fuselage and under each engine nacelle.

History: Up to the beginning of the Second World War, little success had been achieved with Soviet-designed helicopters. Then, in 1939, Professor Bratukhin evolved the Omega. The first prototype was tested in 1941 by K. I. Ponomaryev. The war halted helicopter development to such an extent that no example of the Omega was shown in public until the 1946 Soviet Aviation Display. Although the type was built in limited quantity, it was not widely used. Instead, it served as a scale model for a larger variant. This Omega development was first seen in public at the 1948 Tushino Display and, for his achievements, Professor Bratukhin was awarded the Stalin Prize. The new helicopter was powered by two 550 hp AI-26 radial engines, had the engine and rotor supporting structures replaced by small tapered wings, was fitted with a redesigned glazed nose section, and carried a crew of two and six passengers.

(UK) BRISTOL T.188

First flight: 14 April 1962
Purpose: Supersonic research aircraft, designed to investigate prolonged flight at speeds of up to Mach 3
Power plant: Two Bristol Siddeley (D.H.) Gyron Junior DGJ.10R turbojet engines (each 14,000 lb; 6,350 kg st with reheat)
Wing span: 35 ft 1 in (10.69 m)
Length (excluding nose probe): 71 ft 0 in (21.64 m)
Height: 13 ft 4 in (4.06 m)
Wing area: 396 sq ft (36.79 m^2)
Max level speed: over 1,042 knots (1,200 mph; 1,931 km/h)
Accommodation: Pilot only
Special design features: Stainless steel construction. Modified biconvex wing section, with thickness/chord ratio of 4%. Constant-chord centre-section inboard of nacelles. Leading-edge sweepback outboard of nacelles 38°. Tip of wing, formed by balance area of aileron forward of hinge, had leading-edge sweepback of 64°. Hydraulically-powered ailerons extended from engine nacelles to tips, and comprised in effect all-moving wingtips integral with trailing-edge flap sections. Maximum depth of fuselage 4 ft 11½ in (1.51 m). Maximum width of fuselage 3 ft 9 in (1.14 m). Hydraulically-actuated cascade-type airbrakes on each side of rear fuselage.
History: The T.188 was built under contract from the Ministry of Aviation. Unlike most research aircraft in its class, it was designed to be flown conventionally and to take off on a normal runway. Experience gained in its development, structurally, aerodynamically and in the evolution of its systems, was thought to be able to contribute to the design of future supersonic transport aircraft. Entirely new techniques had to be evolved for the fabrication of the T.188 in welded stainless steel, to meet the problems of kinetic heating. The basic material used was Firth Vickers F.V. 520 stainless steel, able to tolerate a maximum temperature of 500°. In its original form, the aircraft was intended not only for structural and aerodynamic research, but also for investigations into armaments and other propulsion systems. Some of these requirements were subsequently modified to conform with changes in British defence policy. The first T.188 left the factory on 26 April 1961. The second aircraft flew for the first time on 29 April 1963. A third airframe underwent structural testing at the RAE Farnborough from May 1960. No results of the flight tests have been published although it is known that the aircraft's research programme was cut short because of the high rate of fuel consumption at maximum speed.

(USA)

CHANCE VOUGHT XF5U-1

First flight (V-173): 23 November 1942
Purpose: Experimental fighter
Photo: V-173
Power plant: Two Pratt and Whitney R-2000 Twin Wasp engines (each 1,350 hp)
Estimated max speed: 369 knots (425 mph; 683 km/h)
Estimated landing speed: 35 knots (40 mph; 64 km/h)
Special design features and History: The XF5U-1 was an experimental aircraft, with a wing of roughly circular planform which constituted the main structure. The engines were buried in the wing, one on each side of the cockpit, and drove Chance Vought-designed four-blade tractor propellers mounted at the extremities of the wing, through a right-angle transmission with a reduction gearing of approximately 5:1. Special clutches permitted either engine to drive both propellers in an emergency. The propellers were specially developed for this aircraft, and had articulated blades similar to those used on a helicopter, so that at high angles of attack the blades moved forward at constant pitch and flattened out as they moved aft, thus making it possible for the aircraft to hover at low or zero speeds. The purpose of the design was to provide a fighter with a wide speed range. Vertical tail surfaces of the XF5U consisted of twin fins and rudders mounted at the trailing-edge of the wing. The sweptback tailplane and elevators were attached outboard of the vertical surfaces. A low-powered full-scale version of the XF5U-1, of wood and fabric construction, known as the V-173, flew in November 1942 and was fitted with a fixed tailwheel-type undercarriage with spatted main wheels. The XF5U-1 itself was prepared for experimental testing at the USAF base at Muroc (now Edwards), California, in 1947 but never flew.

(UK)

CIERVA W.9

First flight: about 1947
Purpose: Experimental helicopter
Power plant: One de Havilland Gipsy Six Series III six-cylinder piston engine (200 hp)
Accommodation: Crew of two
Special design features: Jet thrust used for torque compensation instead of conventional tail rotor. (A multi-blade fan with variable-pitch blades was used to cool the engine; the air therefrom, after leaving the engine bay, was ducted along the fuselage and heated by mixture with the exhaust gases from the engine. It was then ejected from the port side of the fuselage at the extreme stern. Two horizontal shutters on the outlet controlled the flow, and the pitch of the fan blades was varied by the rudder bar. At normal setting the jet thrust was just sufficient to balance the torque reaction). Metal fuselage of tapered cylindrical form, with a steel-tube structure attached to forward end, carrying main rotor pylon. Vertical fin and rudder. Tailwheel-type landing gear.

History: The Cierva company, formed on 24 March 1926, specialised before the Second World War in the design of various models of the Autogiro, the gyroplane invented by the late Senor Juan de la Cierva. Both the C.30A and C.40 Autogiros were used in small numbers by the RAF. After that time the company conducted experiments in helicopter design, and one of the resulting models was the W.9, a joint product of the Cierva company and G & J Weir Ltd.

(UK)

CIERVA AIR HORSE

First flight: 8 December 1948
Purpose: Experimental helicopter with passenger, freight and crop spraying applications
Power plant: One Rolls-Royce Merlin 24 liquid-cooled piston engine (1,620 hp)
Rotor diameter: 47 ft 0 in (14.33 m)
Length overall: 88 ft 7 in (27.00 m)
Freight compartment: 19 ft x 7 ft 7 in x 5 ft 9 in (5.79 m x 2.31 m x 1.75 m)
Weight empty: 12,140 lb (5,506 kg)
Weight loaded: 17,500 lb (7,938 kg)
Max level speed at S/L: 122 knots (140 mph; 225 km/h)
Max rate of climb at S/L: 1,210 ft/min (369 m/min)
Service ceiling: 23,300 ft (7,100 m)
Max still air range: 286 nm (330 miles; 531 km)
Accommodation: Crew of two or four. Provision for 24 passengers or freight
Special design features: Unique triple-rotor configuration; rotors carried on triangulated outrigger system. All rotors driven in same direction through transmission drives from single distributor gearbox powered by engine. Controllable tail-fins. Wide-track tricycle-type landing gear, each leg comprising an oleo-pneumatic shock-strut with a vertical travel of 5 ft (1.52 m).
Control: Operation of pilot's pitch lever controlled movement of the "collective" exchange spider mechanism in fuselage, giving common collective pitch-change on all three rotors. Pilot's control column provided differential control of rotors in collective pitch.
History: The Air Horse was one of the largest single-engined helicopters in the world at the time of its first flight at Eastleigh, Southampton, piloted by Mr H. A. Marsh. It had been designed originally to meet the requirements of Pest Control Ltd for a large crop-dusting helicopter, but was also capable of accommodating passengers or freight. Two prototypes were constructed to Ministry of Supply contract, and on 12 October 1949 the first of these completed its trials at maximum load, the all-up weight of 17,500 lb being the greatest at which any helicopter had flown up to that time. Following negotiations with the Cierva Autogiro Company, the Saunders-Roe Helicopter Division was formed on 22 January 1951, to take over the premises and current design commitments of the former company. The Cierva technical staff joined Saunders-Roe, and this helicopter became known as the Saro-Cierva Air Horse. The first prototype was lost in an accident in which H. A. Marsh lost his life.

(USA)

CONVAIR SEA DART

First flight: 9 April 1953
Purpose: Experimental delta-winged fighter seaplane
Power plant: XF2Y-1; two Westinghouse J34-WE-42 turbojet engines (each 3,400 lb; 1,542 kg st)
YF2Y-1; two Westinghouse J46 turbojet engines (each 6,000 lb; 2,720 kg st with afterburning)
Wing span: 30 ft 6 in (9.30 m)
Length: 41 ft 2 in (12.55 m)
Height (on hydroskis): 21 ft 1 in (6.43 m)
Weight loaded: about 22,000 lb (9,980 kg)
Max speed in shallow dive at 34,000 ft (10,365 m): over Mach 1 (YF2Y-1)
Take-off run: 5,500 ft (1,676 m)
Landing run: 1,000 ft (305 m)
Accommodation: Pilot only
Special design features: Constructed to float in the water prior to and during initial phase of take-off. Delta wings. Large-area vertical tail. Air inlets for engines on fuselage, above wings, with jet efflux orifices in upper part of tail. Retractable hydroski or skis (the skis gave sufficient hydrodynamic lift on reaching a certain speed to lift the aircraft above the water, where it aquaplaned until take-off).
History: The Sea Dart was developed and built for the US Navy Bureau of Aeronautics. It was the first combat-type aircraft to be equipped with retractable hydroskis, the first delta-winged seaplane, and (YF2Y-1) the first seaplane to exceed Mach 1. The initial aircraft, the XF2Y-1, was first launched in San Diego Bay on 16 December 1952. After a period spent in taxiing trials on water, the first flight was made in April 1953. The XF2Y-1 was followed in the research programme by the more powerful YF2Y-1, which had a modified rear fuselage to accommodate the engine afterburners. On 3 August 1954, the YF2Y-1 exceeded Mach 1 in a shallow dive, but was destroyed in an accident in November 1954. The XF2Y-1 was joined later by three new Sea Darts, built to continue the research programme with both single and twin skis. Only one of these flew, the others having been put in store.

(USA)

CONVAIR XB-46

First flight: 2 April 1947
Purpose: Experimental medium bomber
Power plant: Four Allison-built General Electric TG 180 (J35) turbojet engines (each 4,000 lb; 1,814 kg st)
Wing span: 113 ft 0 in (34.44 m)
Length: 106 ft 0 in (32.31 m)
Height: 28 ft 0 in (8.53 m)
Weight loaded: 91,000 lb (41,277 kg)
Max level speed: about 491 knots (565 mph; 909 km/h)
Service ceiling: about 43,000 ft (13,105 m)
Range with 8,000 lb (3,628 kg) of bombs: about 2,171 nm (2,500 miles; 4,023 km)
Bomb load: 20,000 lb (9,072 kg)
Accommodation: Crew of three
Special design features: Thin straight wings of high aspect ratio. Slim oval-section fuselage. Engines mounted in pairs in underwing nacelles. Main wheels of landing gear retracted forward into the underwing nacelles, between the jet engines.

History: The XB-46 was the first multi-jet bomber built by the Consolidated Vultee Aircraft Corporation. It completed its Phase I (manufacturer's) trials in late 1947, after which it was handed over to the USAF for further tests. In the course of its delivery flight from California to Wright Field, Dayton, Ohio, the XB-46 flew from Oklahoma City to Wright Field at an average speed of 533 mph. However, the six-engined swept-wing Boeing B-47 Stratojet was selected for the USAF, and the XB-46 remained only a prototype.

(USA)

CONVAIR XF-92A

First flight: 18 September 1948
Purpose: Flying mockup to test the delta wing configuration, as a phase in the development of the projected XF-92 jet and rocket powered fighter
Power plant: One Allison J33-A-23 turbojet engine (4,600 lb; 2,086 kg st, or 5,400 lb; 2,450 kg st with water injection). Later replaced by one J33-A-29 turbojet engine, with afterburner (8,200 lb; 3,720 kg st with afterburning)
Wing span: 31 ft 3 in (9.53 m)
Length: 42 ft 5 in (12.93 m)
Height: 17 ft 8 in (5.38 m)
Weight loaded: 15,000 lb (6,804 kg)
Max level speed above 45,000 ft (13,715 m): Mach 0.95
Accommodation: Pilot only
Special design features: Small delta wings with 60° sweepback on the leading-edges and with thickness/chord ratio of 6½%. Elevons fitted on the straight trailing-edges. Large vertical fin and rudder. Circular-section fuselage.
History: The Model 7002 research aircraft was built originally as a flying mockup for the XF-92 fighter and was the first true delta-winged aircraft to fly. However, the contract for the XF-92 was cancelled, after which the Model 7002 was designated XF-92A. Designed with the help of Dr Alexander Lippisch, who had undertaken similar research in Germany, the XF-92A was powered originally by a J33-A-23 engine. In 1951, this power plant was changed for the J33-A-29 and, so powered, the aircraft achieved very high subsonic speeds. In 1952, after it had completed its US Air Force evaluation programme, the XF-92A was handed over to the NACA for further research. Data from the XF-92A was used in the design of the F-102 Delta Dagger fighter, the first prototype of which flew in October 1953, having been designed so far as possible as a 1.22:1 scale-up of the earlier research aircraft.

(USA)

CONVAIR XFY-1

First flight: 2 August 1954
Purpose: Experimental VTOL (vertical take-off and landing) fighter
Power plant: One Allison YT40-A-14 turboprop engine (5,850 ehp)
Wing span: 25 ft 8 in (7.82 m)
Length: 30 ft 9 in (9.37 m)
Tail span: 22 ft 7 in (6.88 m)
Max level speed: about 434 knots (500 mph; 805 km/h)
Accommodation: Pilot only
Special design features: Delta wings, swept at 52° at leading-edge. Very large fin and rudder, and corresponding ventral fin (Jettisonable in emergency horizontal landing). Castor-wheels at apices of all four surfaces to form landing gear. Short stubby fuselage. Pilot's seat mounted on gimbals enabling it to tilt at 45° when aircraft was in vertical position, and to assume normal position for horizontal flight. 16 ft (4.88 m) diameter Curtiss-Wright Turbo-electric co-axial contra-rotating propellers.

History: Like the Lockheed XFV-1, the Convair XFY-1 was conceived for a US Navy design competition held in 1950. It was intended that such an aircraft could be used as an escort fighter able to take-off, fight and land without the need for a carrier deck. A special mobile trolley for transporting the aircraft, and for lowering and raising it for maintenance, was designed by Convair, as was a mobile 20 ft access ladder to enable the pilot to reach his cockpit. Extensive tethered tests were made in a special rig installed in a 195 ft high naval airship hangar at Moffett Field. This rig permitted powered "vertical taxi" tests to be made without the aircraft being beyond control from the ground. In all, some 280 flights were made in this rig before the aircraft was taken out for free flight testing. The XFY-1 made its first free vertical take-off and landing in August 1954. This was followed by another 70 free vertical "up and down" flights before it made its first transition from vertical to horizontal flight and back to the vertical for landing, on 2 November 1954. Development of the type was abandoned subsequently.

(USA) CURTISS-WRIGHT X-100 and MODEL 200/X-19A

First flight (X-100, STOL flight): March 1960
Purpose: VTOL research aircraft
Photo: X-19A
Data: X-100
Power plant: One Lycoming YT53-L-1 shaft-turbine engine (825 shp)
Span, over propellers: 25 ft 0 in (7.62 m)
Length: 28 ft 3½ in (8.63 m)
Height, over propellers: 10 ft 9 in (3.28 m)
Max T-O weight: 3,729 lb (1,692 kg)
Max speed at 5,000 ft (1,525 m): 208 knots (240 mph; 386 km/h)
Special design features: Welded steel-tube fuselage, fabric covered. Short-span wings. Engine drove two Curtiss-Wright hydro-mechanical three-blade propellers, each of 10 ft (3.05 m) diameter. Jet efflux from engine exhausted through nozzles at tail, to provide pitch and yaw control in vertical and low-speed flight. Control in roll was achieved by differential variation of propeller pitch. Four-wheel fixed landing gear.
History: The X-100 was built to develop Curtiss-Wright's "radial lift force" propeller concept. The special propellers were carried on wingtip nacelles, which were tilted by actuating a stick-mounted tilting switch. For vertical take-off landing, the nacelles were roated upward through 90°, so that the propellers functioned as helicopter rotors. Curtiss-Wright claimed that the propellers continued to develop a lift force during forward flight, when they had been tilted down to a conventional position, so that only very small fixed wings were required. Following its first STOL flight, the X-100 completed successful transitions from vertical to horizontal flight and vice versa, and underwent evaluation at NASA's Langley Research Center from October 1960 to October 1961. The X-19A (Model 200) was developed from the X-100. Two X-19As were ordered by USAF Aeronautical Systems Command, the first appearing publicly on 23 July 1963. Development was under the management of the USAF, although the US Army and Navy also evaluated the aircraft. The first X-19A made its maiden hovering flight on 26 June 1964. It was a six-seat twin-engined tandem high-wing aircraft, with four "radial lift force" propellers mounted in tilting wingtip nacelles. The engines were 2,200 shp Lycoming T55-L-5s and the aircraft was of all-metal semi-monocoque construction with a retractable landing gear. Development of the X-19A was terminated in the first half of 1966.

(France)

DASSAULT BALZAC V-001 and MIRAGE III-V

First flight (V-001): 12 October 1962
Purpose: Jet-lift VTOL research aircraft and fighter prototype
Photo: III-V 02
Data: First prototype III-V
Power plant: One SNECMA TF-104 turbofan engine (13,890 lb; 6,300 kg st); subsequently replaced by one SNECMA TF-106 turbofan engine (16,755 lb; 7,600 kg st with afterburning). Eight Rolls-Royce RB.162-1 turbojet engines (each 3,525 lb; 1,600 kg st) for jet-lift
Wing span: 28 ft 7¼ in (8.72 m)
Length overall: 59 ft 0½ in (18.00 m)
Max T-O weight: 29,630 lb (13,440 kg)
Max level speed: Mach 2.04 on second prototype with Pratt and Whitney engine
Accommodation: Pilot only
Special design features and History: The Balzac V-001 jet-lift research aircraft was built to study the problems of vertical flight and to develop the control system designed for the Mirage III-V VTOL fighter. Converted from the original prototype Mirage III airframe, under French government contract, it was powered by a Bristol Siddeley Orpheus turbojet engine (4,850 lb; 2,200 kg st) for propulsion. This engine installation made room for the eight vertically-mounted Rolls-Royce RB.108 engines used for VTOL flight. The first tethered flight was made on 12 October 1962, the first free vertical flight on 18 October 1962, and the first transition on 18 March 1963. This aircraft crashed on 10 January 1964 but was rebuilt for further tests. The Mirage III-V was developed from the V-001 and closely followed the Mirage III-E in general layout. However, the fuselage was lengthened and the wing leading-edges were given compound sweep by increasing the chord near the roots. The first prototype III-V began hovering trials on 12 February 1965. Following replacement of the original engine with the TF-106, the additional power permitted extension of flight tests to forward speeds of up to Mach 1.35 at high altitude and transition from horizontal to vertical flight. The first transition was made on 24 March 1966. Jet-lift was provided by eight engines mounted in pairs in the centre-fuselage, fore and aft of the wheel-wells on each side of the central air duct of the TF-106. Each pair of engines was covered by a rear-hinged sprung-grille intake panel. A bleed-air stabilisation system was employed, with duplicated outlet nozzles under the nose, rear fuselage and outer wings. The second prototype was fitted with a Pratt and Whitney TF30 turbofan engine (18,520 lb; 8,400 kg st with afterburning) but was destroyed in an accident on 28th November 1966. No production was undertaken.

(UK)

DE HAVILLAND DH 108

First flight: 15 May 1946
Purpose: To investigate stability and control problems which arise in aircraft with sweptback wings, and to provide aerodynamic data for the de Havilland DH 106 Comet transport aircraft
Data: Third aircraft
Power plant: One de Havilland Goblin 4 turbojet engine (3,750 lb; 1,700 kg st)
Wing span: 39 ft 0 in (11.89 m)
Length: 26 ft 9½ in (8.17 m)
Wing area: 328 sq ft (30.47 m²)
Accommodation: Pilot only
Special design features: Tapered wings, sharply swept on both leading- and trailing-edges. Standard de Havilland Vampire fuselage, with only slight modifications.
History: The de Havilland DH 108 was a tail-less experimental monoplane of which three examples were built for two main purposes. Initial work on the project began in October 1945 and, to get the aircraft completed as quickly as possible, a standard Vampire fuselage was used, complete with Goblin engine. The first aircraft (TG283) was designed to determine the low-speed characteristics of the swept wing and was fitted with fixed open wing-slots. The second prototype (TG306) followed. This aircraft was built to assess the high-speed characteristics of the wing and was fitted with retractable wing-slots. It was in this aircraft that Geoffrey de Havilland Jr was killed in September 1946. A third aircraft (VW120) was then built to continue the high-speed programme initiated with the second. Fitted with a Goblin engine of higher rating, this aircraft embodied a number of structural changes, had power-assisted controls, a sharper nose, lower cockpit with armoured canopy, and ejector seat. On 12 April 1948, piloted by John Derry, VW120 established an international speed record over a 100 km closed circuit of 525.59 knots (605.23 mph; 973.81 km/h). On 6 September 1948, the same aeroplane, with the same pilot, exceeded the speed of sound in a dive from 40,000 ft to 30,000 ft, recording a Mach number between 1.0-1.1. This was the first British aircraft and the first jet-powered aircraft to achieve a speed of over Mach 1.

(Canada/USA) DE HAVILLAND CANADA XC-8A/BELL AIR CUSHION LANDING SYSTEM

First flight (take-off with ACLS): 31 March 1975
Purpose: Research into air cushion landing systems
Power plant, dimensions, etc: Basically as for de Havilland Canada DHC-5 Buffalo utility transport aircraft, except for modifications as noted below (see Pocket Book 5)

Special design features and History: Bell Aerospace began development of an air cushion landing system (ACLS) as a company-funded research project in December 1963. In 1966 it received a $99,000 contract from the USAF Flight Dynamics Laboratory for wind-tunnel testing of the project. Subsequent Air Force contracts were placed for a $99,500 feasibility study in 1966, a $98,700 model test programme in 1967 and a $66,300 flight test programme in 1968.

The initial intention was to determine the best form of ACLS for cargo transports, and the first flight test programme was carried out with a modified Lake LA-4 amphibian. Current activities are funded under a joint United States/Canadian programme aimed at adapting the ACLS for military transport aircraft. This would allow such aircraft to operate from a variety of surfaces, including rough fields, soft soils, swamps, water, ice and snow. It was decided to use a de Havilland Canada XC-8A Buffalo STOL military transport aircraft as a testbed for this programme.

The ACLS is based on the ground effect principle that employs a layer of air instead of wheels as an aircraft's ground contacting medium. The system's trunk, a large inner-tube-like arrangement, encircles the underside of the fuselage. Upon inflation, the trunk provides an air duct and seal for the air cushion. The cushion air pressure is provided by an on-board auxiliary compressor, and the underside of the rubberised trunk is perforated with hundreds of vent holes through which the air is allowed to escape to form the air cushion. Balancer floats have also been mounted on struts beneath each wing for operation on or from water. Beneath the floats are sprung skids for use in operation from land. Pratt and Whitney (Canada) was made responsible for development and flight qualification of the auxiliary power system, and de Havilland Aircraft of Canada modified the XC-8A to take the ACLS installation. At the beginning of 1975 the USAF was conducting flight tests of the ACLS aircraft at Wright-Patterson AFB, Dayton, Ohio.

(USA)

DOAK VZ-4DA

First flight: 25 February 1958
Purpose: VTOL convertiplane research aircraft
Power plant: One Lycoming YT53 shaft-turbine engine (840 shp)
Wing span (over ducts): 25 ft 6 in (7.77 m)
Length: 32 ft 0 in (9.75 m)
Height: 10 ft 0 in (3.05 m)
Wing area: 96 sq ft (8.92 m^2)
Weight loaded: 3,200 lb (1,452 kg)
Max level speed at S/L: 200 knots (230 mph; 370 km/h)
Rate of climb at S/L: 4,000 ft/min (1,220 m/min)
Hovering ceiling: 6,000 ft (1,830 m)
Range: 200 nm (230 miles; 370 km)
Accommodation: Crew of two
Special design features and History: Developed under contract from the US Army Transportation Research and Engineering Command, the Doak Model 16 (military designation VZ-4DA), was a simply-constructed research aircraft with ducted propellers, or fans, on its wingtips. The ducts could be rotated through 90°, enabling the propellers to function as helicopter rotors for vertical take-off, landing and hovering. At a safe height after take-off, the ducts were rotated to the normal forward-facing attitude, making possible much higher cruising speeds than with a conventional helicopter. Duct rotation was initiated by means of a switch on the control column. The fuselage remained horizontal at all times, and the Model 16 required no automatic flight controls, automatic stabilisation, artificial damping or power boost. Construction was conventional, with a welded steel-tube fuselage, covered with moulded plastic skin. The engine was mounted in the fuselage behind the tandem crew seats, and was fitted with a long tailpipe. Horizontal and vertical vanes of stainless steel were hinged at the jet efflux, to assist control during low-speed and vertical flight. Each duct contained 14 glassfibre inlet guide vanes, which were deflected to provide roll control during hovering. The propellers had eight fixed-pitch stainless steel blades. Nine stator blades were located aft of the propeller in each duct. First flight tests took place at Torrance Municipal Airport. On completion of the makers' trials, the aircraft was transferred to Edwards Air Force Base for further testing. As a result of its successful 50-hour flight-testing programme at the Base, the Model 16 was accepted by the US Army in September 1959, for further Army/NASA evaluation.

(Germany)

DORNIER Do 31 E

First flight: 10 February 1967
Purpose: Experimental V/STOL transport aircraft
Photo: Do 31 E 3
Power plant: Two Rolls-Royce (Bristol) Pegasus 5-2 vectored-thrust propulsion engines (each 15,500 lb; 7,000 kg st). Removable lift-jet pod on each wingtip, each housing four Rolls-Royce RB.162-4D turbojet engines (each 4,400 lb; 2,000 kg st)
Wing span: 59 ft 3 in (18.06 m)
Length overall: 68 ft 6 in (20.88 m)
Height: 28 ft 0 in (8.53 m)
Max payload: 11,000 lb (4,990 kg)
Max T-O weight: 60,500 lb (27,442 kg)
Cruising speed at 20,000 ft (6,100 m): 347 knots (400 mph; 650 km/h)
Service ceiling: 34,500 ft (10,515 m)
Accommodation: Crew of two, plus provision for 34 troops, 24 stretchers or freight
Special design features: Wings with no dihedral or incidence, swept-back at 8°30′ at quarter-chord. Two-section ailerons, on outer wings, took form of camber-changing flaps with movement of ±25°. Circular-section fuselage with rear-loading doors. Swept vertical tail and fixed-incidence tailplane mid-set on fin. Main engines in pods under wings. Lift-jets mounted in line astern and fitted with deflector nozzles.
History: Two prototypes were constructed under a German Defence Ministry contract, with the assistance of Vereinigte Flugtechnische Werke and Hamburger Flugzeugbau. The first, the Do 31 E 1, flew in February 1967, followed by the second, the Do 31 E 3, on 14 July 1967. The first transition from vertical take-off to horizontal flight, by E 3, took place on 16 December 1967, and the first transition from horizontal flight to vertical landing on 21 December. A third airframe, the Do 31 E 2, was used for static tests. During 1969-70, the Do 31 E was studied by a team of scientists and engineers from NASA. By mid-1969, 200 tests, including 110 take-offs and many transitions, had been made, and testing continued until the programme was terminated in April 1970.

(USA)

DOUGLAS D-558-1 SKYSTREAK

First flight: 28 May 1947
Purpose: To obtain in free flight air-load measurements not obtainable in wind-tunnels of the time
Power plant: One Allison J35-A-23 turbojet engine (4,000 lb; 1,814 kg st), replaced later by a J35-A-11 turbojet engine (5,000 lb; 2,268 kg st)
Wing span: 25 ft 0 in (7.62 m)
Length: 35 ft 1 in (10.69 m)
Height: 12 ft 2 in (3.71 m)
Wing area: 150 sq ft (13.94 m^2)
Weight loaded: 9,750 lb (4,423 kg)
Max level speed: 565 knots (651 mph; 1,047 km/h)
Accommodation: Pilot only
Special design features: NACA laminar-flow wing section with an aspect ratio of 4.15 and a 10% thickness/chord ratio. Circular-section fuselage, with the rear section constructed as a thick-gauge magnesium alloy tube. Nose section, from rear of pilot's seat, jettisonable in an emergency. Automatic pressure-recording system fitted, with 400 measurement points on the fuselage, wings and tail. Strain gauges fitted to selected points on wings and tail.

History: Conceived in 1945, the Skystreak was designed by the Douglas company for the US Navy Bureau of Aeronautics, in conjunction with the NACA. The first of three Skystreaks made its maiden flight in May 1947 at Muroc Dry Lake. Less than three months later, on 20 August, this aircraft set up a World Speed Record of 640.74 mph (1030.95 km/h). Five days later the same aircraft raised the record to 650.92 mph (1,047.33 km/h). The Skystreak flown on the record flights had a maximum take-off weight of over 10,000 lb (4,536 kg), and carried approximately 230 US gallons of fuel and 640 lb (290 kg) of special instrument equipment in addition to the normal complement of pilot's instruments and gear.

(USA)

DOUGLAS D-558-2 SKYROCKET

First flight: 4 February 1948
Purpose: Investigation of sweptback wings
Power plant: One Westinghouse J34-WE-22 turbojet engine (3,000 lb; 1,360 kg st) and one Reaction Motors XLR-8 bi-propellant rocket motor (6,000 lb; 2,721 kg st)
Wing span: 25 ft 0 in (7.62 m)
Length: 45 ft 3 in (13.79 m)
Height: 11 ft 6 in (3.51 m)
Max speed: Mach 2.005
Service ceiling: over 83,200 ft (25,360 m)
Accommodation: Pilot only
Special design features: Aerofoil sections of conventional subsonic type with rounded leading-edges and contours. 35° wing sweepback at quarter-chord. 40° tailplane sweepback at same location. Handley Page automatic leading-edge slots. Turbojet air intake in the underside of the fuselage. Rocket motor in the aft end of the fuselage. Nose of the fuselage, containing the pilot's cockpit, jettisonable in an emergency.
History: Shortly after the Skystreak research project was started, the contract was modified to include investigation of sweptback wings in addition to the straight wings of the Skystreak. As a full realisation of the advantages of the swept wing could not be obtained with the limited power of turbojet engines then available, it was necessary to fit a supplementary rocket motor for use during take-off and high-speed flight. The redesigned aircraft was named Skyrocket, and three were built. In August 1951, one of them, with its jet engine removed and its fuel capacity doubled, attained a speed of 1,075 knots (1,238 mph; 1,992 km/h); a few days later it reached an altitude of 79,494 ft (24,230 m). On both occasions the Skyrocket was carried to 35,000 ft (10,670 m) under a Boeing P2B-1 aircraft. On 31 August 1953, a Skyrocket climbed to 83,235 ft (25,370 m) and, on 20 November of the same year, one accelerated to Mach 2.005 after being dropped from the mother plane at 32,000 ft (9,755 m). This was the first time that a piloted aircraft had flown at twice the speed of sound. The Skyrockets completed further research with NACA, at the High Speed Research Station at Muroc Dry Lake (now Edwards AFB), California, totalling 161 flights of which 74 were combined jet and rocket-powered.

(USA)

DOUGLAS X-3

First flight: 20 October 1952
Purpose: Research into the efficiency of turbojet engines and short-span double-wedge wings and tailplanes, and to collect information on thermodynamic heating at high altitudes and at speeds of up to Mach 3
Power plant: Two Westinghouse J34-WE-17 turbojet engines (each 4,200 lb; 1,905 kg st with afterburning)
Wing span: 22 ft 8 in (6.91 m)
Length: 66 ft 9 in (20.35 m)
Height: 12 ft 6 in (3.81 m)
Weight loaded: 27,000 lb (12,247 kg)
Max designed speed: Mach 3
Accommodation: Pilot only
Special design features: Very short-span wings with a thickness/chord ratio of 5%. Wing leading-edge flaps and split trailing-edge flaps. Long slim fuselage with pointed nose and a thin tailboom, on which were mounted small tail surfaces. One intake for engines on each side of fuselage, immediately aft of pilot's cabin; jet exhausted under tailboom. 1,200 lb (544 kg) of research instruments. Over 850 "pin-hole" orifices which recorded pressures over various portions of airframe. Temperature readings registered at 150 points; stresses and air loads indicated by 185 strain gauges.
History: The X-3 was designed and built by the Santa Monica Division of Douglas, under the joint sponsorship of the USAF, the US Navy and NACA. The project was directed by the Air Research and Development Command, USAF, to test design features of an aircraft suitable for sustained flights at extremely high altitudes. The contract for the X-3 was placed with Douglas by the USAF in 1947. The mock-up of the aircraft was completed in 1948, and construction began in 1949. Design and construction presented problems of unprecedented complexity, involving not only aerodynamics but the use of then new materials and methods. These included the development of fabrication and construction techniques with titanium, used extensively throughout the X-3. The aircraft was flown for the first time at Edwards Air Force Base, California. After a period of testing in the hands of Bill Bridgeman, chief test pilot of the Douglas Company, the X-3 was turned over to NACA for further research in late 1953. After twenty flights had been made by NACA, the programme was terminated on 23 May 1956.

(USA)

DOUGLAS XB-42 and XB-43

First flight (XB-42): June 1944
Purpose: Experimental medium attack bomber
Photo: XB-42
Data: XB-42
Power plant: Two Allison V-1710-125 piston engines (each 1,725 hp)
Wing span: 70 ft 0 in (21.34 m)
Length: 53 ft 0 in (16.15 m)
Weight empty: 19,149 lb (8,686 kg)
Weight loaded: 35,702 lb (16,194 kg)
Max level speed: 356 knots (410 mph; 660 km/h)
Range: about 4,342 nm (5,000 miles; 8,047 km)
Armament: Four 0.5-in remotely-controlled and sighted guns in trailing-edge of wings, firing aft. Attack version intended to have nose section containing various combinations of guns, ranging from eight 0.5-in to one 75 mm cannon and two 0.5-in guns. Maximum bomb load 8,000 lb (3,628 kg), carried internally.
Accommodation: Crew of three
Special design features: Straight wing of laminar-flow aerofoil section. All metal ailerons and slotted trailing-edge flaps. Cruciform tail surfaces; twin fins and rudders, one above and one below fuselage. Engines drove contra-rotating pusher propellers positioned behind rudders, through extension shafts. Tricycle landing gear with emergency bumper in bottom of lower fin.
History: The XB-42 was an experimental piston-engined bomber developed by Douglas to US Air Technical Service Command contract. On 8 December 1944, the first prototype flew from Long Beach, California, to Washington, DC, a distance of 2,290 miles (3,685 km), at an average speed of 432 mph (695 km/h). This aircraft crashed at Washington on 15 December 1945, through no fault of either airframe or power plant installation. The second prototype differed from the original model by having a single cockpit enclosure in place of the latter's distinctive twin blister canopies. The XB-43 was generally similar to the XB-42 except for having a slightly greater wing span, reduced length, two General Electric TG-180 turbojet engines (each 4,000 lb; 1,814 kg st) and a conventional tail unit. Two examples were built, becoming the USAF's first jet bombers.

(Germany)

EWR VJ 101C

First flight (first free hover): 10 April 1963
Purpose: Experimental VTOL aircraft
Power plant (X-1): Six Rolls-Royce RB.145 engines (2,750 lb; 1,250 kg st dry). The X-2 had afterburning engines (each 3,550 lb; 1,610 kg st) in the wingtip pods
Wing span: 21 ft 8½ in (6.61 m)
Length overall: 51 ft 6 in (15.70 m)
Height over tail: 13 ft 6½ in (4.13 m)
Max T-O weight for VTOL (X-2): 17,635 lb (8,000 kg)
Max speed achieved (X-1): Mach 1.08
Accommodation: Pilot only
Special design features and History: Two examples of this aircraft were built and tested, to provide data to assist the development of a projected Mach 2 VTOL fighter. The RB.145 engines were developed jointly by Rolls-Royce and MAN-Turbomotoren GmbH of Germany. The X-1 made its first free hover in April 1963, after initial tests on an extensible pylon. First horizontal take-off was made on 31 August 1963, and first transition on 20 September 1963. Mach 1 was exceeded in level flight on several occasions before the X-1 crashed on 4 September 1964. The X-2 was basically similar to the X-1 but had afterburning engines to make possible speeds well above Mach 1. Pylon tests began in the Spring of 1965 and it made its first hovering flight without reheat on 12 June 1965. On 22 October 1965, the X-2 made its first complete transitions from vertical to horizontal flight and vice versa, using reheat.

In the fuselage engine bay were two vertically-mounted RB.145 lift engines, which were used only in VTOL and low-speed flight. Each swivelling wingtip pod contained two engines. Extensive research was necessary to obtain efficient air intake design. The cowl which covered the fuselage engines during normal flight was raised to a specific angle for VTOL. Air was supplied to the wingtip engines in vertical flight through a slot running round the entire pod and produced by sliding forward the front portion of the pod cowling. As the fuselage engines were forward of the wingtip engines, giving triangulated lift forces, it was possible to control the aircraft in VTOL flight by thrust modulation, so dispensing with the need for "puffer-pipes". Transition from forward flight to touchdown of the VJ 101 took about 90 seconds.

(USA)

FAIRCHILD VZ-5FA

First flight (tethered): 18 November 1959
Purpose: VTOL research aircraft
Power plant: One General Electric YT58-GE-2 shaft-turbine engine (1,024 shp) driving four Hartzell three-blade propellers
Wing span: 32 ft 9 in (9.98 m)
Length: 33 ft 8 in (10.26 m)
Height: 16 ft 10 in (5.13 m)
Gross wing area: 191 sq ft (17.74 m^2)
Area of VTOL flaps: 126 sq ft (11.71 m^2)
Weight loaded for VTOL testing: 3,976 lb (1,803 kg)
Max level speed (design limited) at S/L: 160 knots (184 mph; 297 km/h)
Accommodation: Pilot only
Special design features: Constant-chord wing with no dihedral. Aspect ratio 5.62. Incidence 5°. Wing fitted with full-span VTOL flaps of large area. Braced T-type tail unit; one-piece horizontal surface with variable incidence. Two small four-blade propellers at trailing-edge of tailplane for control in vertical and low-speed flight. Non-retractable tricycle landing gear. Tailskid for VTOL operation.
History: Developed for the US Army, the VZ-5FA, or Fairchild M-224-1, was an experimental vertical take-off and STOL aircraft which employed the deflected slipstream technique. When stationary the VZ-5FA, had its wing flaps retracted and looked like a normal light aircraft, except for having four propellers mounted on the wing leading-edge. In VTOL take-off position, however, the aircraft took on a strange appearance. In this attitude the aircraft poised on its main landing gear and tailskid, with the nose of the fuselage and nosewheel pointing upwards at an angle of about 45°. The wing flaps were extended downwards, so that the slipstream from the propellers would be directed towards the ground. Tethered flight tests began in November 1959.

(USA)

FAIRCHILD XC-120 PACK-PLANE

First flight: 11 August 1950
Purpose: Experimental prototype detachable-fuselage transport aircraft
Power plant: Two Pratt and Whitney R-4360 radial engines (each 3,250 hp)
Wing span: 106 ft 6 in (32.46 m)
Length: about 77 ft 1 in (23.50 m)
Cargo compartment:
> Height 8 ft 0 in (2.44 m)
> Width 9 ft 2 in (2.79 m)
> Length 36 ft 11 in (11.25 m)
> Cargo space volume 2,900 cu ft (82.12 m^3)

Accommodation: Crew of five
Special design features: Wings, power plant, tailbooms and tail unit similar to those of the Fairchild C-119 Packet military transport aircraft. Shallow flat-bottomed nacelle housed crew in forward section; underneath this nacelle was attached the cargo compartment or pack. Clamshell loading doors at front and rear of detachable pack for easy loading. Space between pack and carrier sealed by rubber tube, which could be inflated after pack was secured in place. Pack handling gear consisted of four dual-wheel units which could be attached before pack was lowered to the ground. Pack raised and lowered by four electrically-operated drum hoists, built into the carrier. Release and locking mechanism operated simultaneously or separately at all four fittings by controls in crew nacelle.
History: The USAF placed a contract with Fairchild for an experimental prototype detachable-fuselage transport aircraft, based on the C-119 Packet, in the late 1940s. It was envisaged that various types of packs, fitted for different military functions, could be provided. Preliminary designs called for a pack with a 9-ton payload capacity. The XC-120 could fly with or without the detachable pack, and some attention was given to the possibility of dropping the pack in flight or skidding it onto the ground during flyover at a low height. Although the principle of the Pack-Plane proved successful, no production of the type was undertaken.

(UK)

FAIREY F.D.1

First flight: 12 March 1951
Purpose: Delta-wing research aircraft
Power plant: One Rolls-Royce Derwent 8 turbojet engine (3,600 lb; 1,633 kg st)
Wing span: 19 ft 6½ in (5.96 m)
Length: 26 ft 3 in (8.00 m)
Weight loaded: 6,800 lb (3,084 kg)
Max level speed: 545 knots (628 mph; 1,011 km/h)
Accommodation: Pilot only
Special design features: Horizontal tail surfaces (the first delta-winged aircraft with such surfaces to be flown). Wingtip fairings for anti-spin parachutes. Drogue parachute to reduce landing speed and run, housed in a tail fairing.
History: The F.D.1 was a very small aircraft intended originally as an experimental vertical take-off fighter. After taking-off from a purpose-built inclined ramp, it was to have been controlled in flight by the efflux from four small nozzles that surrounded the main engine orifice. The F.D.1 followed earlier research carried out with un-manned VTO project aircraft from 1949. These rocket-powered aircraft initially proved difficult to control but subsequent development enabled more successful flights to be made. The concept of a vertical take-off fighter was, however, abandoned, and the F.D.1 was fitted out for normal runway operations, without the control nozzles. Taxiing trials were made at Ringway Airport from 12 May 1950. The first flight was made from the RAF Station at Boscombe Down and lasted 17 minutes. The two side-mounted fairings that were to have contained two of the four ducts for the efflux control system were subsequently removed. The F.D.1 proved difficult to control in flight and was involved eventually in a landing accident, after which flying was suspended.

(UK)

FAIREY DELTA 2

First flight: 6 October 1954
Purpose: To investigate the characteristics of flight and control at transonic and supersonic speeds
Power plant: One Rolls-Royce Avon turbojet engine with afterburning (12,000 lb; 5,443 kg st Avon RA.5 in WG774: 13,000 lb; 5,896 kg st Avon RA.28 in WG777)
Wing span: 26 ft 10 in (8.18 m)
Length: 51 ft 7½ in (15.72 m)
Height: 11 ft 0 in (3.35 m)
Wing area: 360 sq ft (33.45 m^2)
Weight loaded: 13,884 lb (6,297 kg)
Max level speed: 996 knots (1,147 mph; 1,846 km/h)
Max range: 721 nm (830 miles; 1,335 km)
Accommodation: Pilot only
Special design features: Delta wing with Fairey symmetrical wing section. Thickness/chord ratio 4%. Aspect ratio 2.0. Chord 25 ft (7.62 m) at root, 13 ft 6 in (4.11 m) mean. Incidence 1.5°. Sweepback 60°. Long pointed fuselage, entire nose section to aft of cockpit able to pivot downward to improve pilot's view during take-off and landing. No tailplane. Swept fin and rudder.
History: Two examples of the Fairey Delta 2 research aircraft were built under contract from the Ministry of Supply. The first aircraft (WG774) made its maiden flight at Boscombe Down, but was damaged during landing on 17 November of the same year. However, on 10 March 1956 this aircraft, flown by L. Peter Twiss, became the first to set an over-1,000 mph world speed record, with an average of two flights of 983 knots (1,132 mph; 1,821 km/h). The second aircraft (WG777) exceeded the speed of sound during its first flight on 15 February 1956. This aircraft was later sent to Cazaux, France, for supersonic flight tests at heights down to 3,500 ft (1,066 m). In all, 52 flights were made in one month from 15 October 1956. Eventually, the first aircraft was converted into the BAC 221 (which see). The drooping nose pioneered on the Delta 2 was adopted subsequently as a feature of the Concorde design.

(UK)

FAIREY GYRODYNE

First flight: 7 December 1947
Purpose: Experimental helicopter
Photo: Jet Gyrodyne
Power plant: One Alvis Leonides radial engine (525 hp)
Rotor diameter: 52 ft 0 in (15.85 m)
Fuselage length: 19 ft 2 in (5.84 m)
Height: 10 ft 1 in (3.07 m)
Weight loaded: 4,800 lb (2,177 kg)
Normal operating speed: 96 knots (110 mph; 177 km/h)
Range: 217 nm (250 miles; 402 km)
Accommodation: Seating for four to five persons
Special design features: Two-blade Fairey variable-pitch propeller located at end of starboard stub-wing, both main rotor and propeller being driven by the Alvis Leonides engine which was mounted within fuselage, aft of cockpit and rotor pylon. Twin fins and rudders. Trim-tab along trailing-edge of tailplane. No elevators. Tricycle landing gear. Cooling air for engine drawn through intake grill in forward face of rotor pylon casing. Air exhausted through opening in fuselage side adjacent to engine exhaust pipes to provide ejector action. Dual controls.

History: The Gyrodyne was a four/five-seat helicopter which incorporated several then new principles in design. The more usual control and anti-torque rotor was dispensed with, and its functions assumed by the conventional propeller located on the stub-wing. The offset tractor propeller also made it possible for the aircraft to be flown as an autogyro; and engine failure did not entail rapid action on the part of the pilot to avoid loss of rotor rpm, because the main rotor blades were always flying within autorotative pitch range. A smaller angle of incidence used in transitional flight resulted in a reduction in axial flow through the rotor in cruising flight, with less likelihood of blade tip stall and reduced vibration. The first Gyrodyne prototype flew untethered in December 1947. On 28 June 1948, it set up an international helicopter speed record of 107.9 knots (124.3 mph; 200 km/h). However, on 17 April 1949, two days before an attempt was to have been made on the 100 km closed-circuit speed record, the aircraft crashed after failure, due to fatigue, of the rotor head. The crew were killed. The second Gyrodyne was modified subsequently into the Jet Gyrodyne, with tip-jet rotor drive

(UK)

FAIREY ROTODYNE

First flight: 6 November 1957
Purpose: Experimental VTOL transport convertiplane
Data: Rotodyne Y prototype
Power plant: Two Napier Eland N.E.L.3 turboprop engines (each 3,000 ehp)
Rotor diameter: 90 ft 0 in (27.43 m)
Wing span: 46 ft 6 in (14.17 m)
Length: 58 ft 8 in (17.88 m)
Height: 22 ft 2 in (6.76 m)
Weight loaded: Flew at weights up to 33,000 lb (14,968 kg)
Max level speed: 161 knots (185 mph; 298 km/h)
Range: 247-391 nm (285-450 miles; 459-724 km)
Accommodation: Pilot and co-pilot, plus 40 passengers
Special design features: Single four-blade main rotor carried above fuselage on massive pylon structure, fully faired early in 1960. Pressure-jet units of Fairey design at rotor-blade tips. Engines mounted in nacelles underslung from high-mounted straight wings, driving four-blade propellers. Ailerons fitted to wings. Three fins and two rudders on tail unit. Retractable tricycle landing gear
History: The Rotodyne was based on experience gained with the Fairey Gyrodyne. On 5 January 1959 the Rotodyne Y, as the prototype was designated, established a world speed record for rotorcraft with a speed of 190.9 mph (307.22 km/h) over a closed circuit of 100 km. The Rotodyne Y was considerably smaller than the intended 54/70-passenger production version (Rotodyne Z) which was to be powered by 5,250 shp Tyne engines. Many airline operators around the world became interested in purchasing production aircraft, as the Rotodyne was the first vertical take-off airliner. In operation the Rotodyne's engines were coupled to auxiliary air compressors through a clutch. The compressors delivered air to pressure-jet units at the rotor-tips, where fuel was burned to produce thrust rotation. The Rotodyne could be flown as a pure helicopter, with tip-jets burning and the propellers giving no thrust, or as an autogyro, with the engine power going only to the forward-facing propellers and the rotor autorotating. In practice, the tip-jets were normally used for take-off and landing, while for cruising flight the rotor autorotated. The first transition from vertical to horizontal flight was achieved on 10 April 1958. The Rotodyne programme was abandoned in 1962 for economic reasons, including the likely high cost of a programme to reduce tip-jet noise levels

(UK)

FAIREY V.T.O. PROJECT

First flight (successful): 1 May 1949
Purpose: To explore the practicability of launching aircraft from short ramps at low accelerations
Power plant: One Fairey-developed Beta I rocket motor with two nozzles (900 lb; 408 kg thrust)
Wing span: 10 ft 0 in (3.05 m)

Special design features and History: First seen publicly at the Society of British Aircraft Constructors (SBAC) Show of 1952, the VTO project was designed to test the basic configuration of the projected Fairey F.D.1 research aircraft with an unmanned vehicle. Details were released in 1953 of one of a series of vertical take-off delta-wing rocket-powered aircraft models which had been tested successfully on the Woomera Rocket Range, Australia, including launches in near-vertical climbs. The Fairey company had been engaged in the development of this project for several years, under a Ministry of Supply contract, before these first details were released, the first of about 40 VTO models having been fired in Cardigan Bay, Wales, in 1949. Each VTO model was powered by a rocket motor with two nozzles, one of which could be swivelled laterally and the other vertically in response to signals from the autopilot. The resulting mean thrust line could thus be varied to maintain controlled flight at low speeds. Two 600 lb (272 kg) solid-fuel booster rockets were also used during initial stages of launching. The early models proved difficult to control because of troubles with the gyro-stabilized autopilot, but this was later put right and further tests proved more successful.

(France)

FOUGA C.M. 88-R GEMEAUX

First flight: 6 March 1951
Purpose: Experimental aircraft, used as testbed for Turboméca turbojet engines
Photo: Gemeaux I
Power plant: See below
Wing span: 35 ft 3¾ in (10.76 m)
Length: 21 ft 10¼ in (6.66 m)
Height: 6 ft 4 in (1.93 m)
Weight loaded: See below
Max level speed (Gemeaux I): 154 knots (177 mph; 285 km/h)
Max level speed (Gemeaux II): 178 knots (205 mph; 330 km/h)
Max level speed (Gemeaux III): 215 knots (248 mph; 400 km/h)
Max level speed (Gemeaux IV): 135 knots (155 mph; 250 km/h)
Max level speed (Gemeaux V): 178 knots (205 mph; 330 km/h)
Service ceiling: 9,850-32,800 ft (3,000-10,000 m)
Accommodation: Crew of two
Special design features: Fuselages and outer wings of two Cyclope aircraft; the fuselages being joined together by a centre-wing and an aft connecting member. Each fuselage had a single cockpit and "butterfly" tail, the inner panels of the latter being shortened so as not to touch. Engine or engines mounted externally on top of the fuselages or on wing centre-section. Four-wheeled landing gear.
History: Two basic Gemeaux airframes were built to serve as testbeds for various types of turbojet engines. These aircraft were flown in the following forms:
Gemeaux I. Powered by two Turboméca Piméné engines (each 220 lb; 100 kg st), one on each fuselage. Weight loaded was 2,416 lb (1,096 kg).
Gemeaux II. Powered by one Turboméca Marboré I (605 lb; 275 kg st), mounted on centre-wing midway between fuselages. First flown on 16 June 1951. Weight loaded was 2,570 lb (1,166 kg).
Gemeaux III. Powered by one Turboméca Marboré II (880 lb; 400 kg st), mounted as in Mk II. First flown with prototype Marboré II (770 lb; 350 kg st) on 24 August 1951 and with production engine on 2 January 1952. Weight loaded was 2,724 lb (1,236 kg).
Gemeaux IV. Powered by one Turboméca Aspin I ducted-fan turbojet (441 lb; 200 kg st). First flown on 6 November 1951. Weight loaded was 2,581 lb (1,171 kg).
Gemeaux V. Powered by one Turboméca Aspin II ducted-fan turbojet (794 lb; 360 kg st). First flown 21 June 1952. Weight loaded was 2,724 lb (1,236 kg).

(UK) HANDLEY PAGE H.P. 88

First flight: 21 June 1951
Purpose: To investigate the flight characteristics of the crescent wing for use in larger form on the Handley Page Victor bomber
Power plant: One Rolls-Royce Nene 102 turbojet engine (5,100 lb; 2,313 kg st)
Wing span: 40 ft 0 in (12.19 m)
Length: 40 ft 0 in (12.19 m)
Weight loaded: 14,640 lb (6,640 kg)
Max level speed: about Mach 0.9
Accommodation: Pilot only

Special design features: One-quarter scale wing of the Victor bomber. The wing had three different degrees of sweepback, and was fitted with outer-section leading-edge flaps and rearward-moving trailing-edge flaps. Supermarine Type 521 (basically a Supermarine Attacker) fuselage. New tail surfaces with an all-moving high-mounted tailplane. Rear-fuselage airbrakes. Tailwheel-type landing gear.

History: Another basic wing configuration that originated in Second World War Germany, the crescent wing became less swept and had a reduced thickness/chord ratio towards its tips. This was to give the wing an unchanging critical Mach number along its whole span. The wing of the H.P.88 was built by Blackburn and General Aircraft in the late 1940s and the completed aircraft was known at first as the Blackburn Y.B.2. The H.P.88, as it later became known, was involved in an accident before the full potential of the new wing had been explored. However, the crescent wing configuration was adopted for the Handley Page Victor bomber, versions of which are still used by the R.A.F. as flight refuelling tankers.

(UK)

HANDLEY PAGE H.P. 115

First flight: 17 August 1961
Purpose: Aerodynamic research aircraft
Power plant: One Bristol Siddeley Viper 9 turbojet engine (1,900 lb; 862 kg st)
Wing span: 20 ft 0 in (6.10 m)
Length overall: 45 ft 0 in (13.72 m)
Wing area: 430 sq ft (39.95 m²)
Max T-O weight: 5,000 lb (2,268 kg)
Max level speed: about 261 knots (300 mph; 483 km/h)
Accommodation: Pilot only
Special design features and History: The H.P. 115 was built to a Ministry of Aviation contract. It had a wing of slender, low aspect ratio (0.925) delta planform, and the engine was mounted above the rear of the fuselage at the base of the tail-fin. Construction was all-metal, except for the rudder and full-span tab-controlled elevons which were fabric-covered.

For its trials the H.P. 115 had a wing with leading-edge sweep of 74° 42' and lacking any form of conical camber or ogival planform; but the leading-edge was detachable to permit flight testing of a wide variety of shapes, if required. A large perforated airbrake was fitted under each wing, ahead of the main legs of the non-retractable tricycle landing gear. The large fairing on the leading-edge of the fin contained a camera to photograph wool tufts on the wing during flight testing. A housing for an anti-spin and braking parachute was situated under the rudder.

The H.P. 115 made possible for the first time practical flying experience with a wing of this type, and provided data for the Concorde airliner. It flew for the first time on 17 August 1961 and was subsequently flown intensively by pilots of the RAE, Bedford, to study stability, control and handling characteristics

(UK)

HANDLEY PAGE H.P.75 MANX

First flight: 24 August 1943
Purpose: To investigate problems associated with tail-less aircraft
Power plant: Two de Havilland Gipsy Major four-cylinder piston engines (each 140 hp)
Span: 40 ft 0 in (12.19 m)
Length: 18 ft 0 in (5.49 m)
Wing area: 246 sq ft (22.85 m²)
Weight loaded: 4,000 lb (1,814 kg)
Cruising speed: 130 knots (150 mph; 241 km/h)
Service ceiling: 15,000 ft (4,570 m)
Accommodation: Crew of two
Special design features: Fuselage was a metal monocoque structure, with pilot accommodated at front and observer in another cabin at rear and facing aft. Mid-positioned wings built in three sections, comprising a constant-chord centre-section and two sweptback and tapered outer wings. Outer sections fitted with elevons, with split trailing-edge flaps between them and the fuselage, and leading-edge slots. Cantilever fins and mass-balanced rudders at extremities of wings. Third fin fitted at rear of fuselage. Engines mounted as pushers, with two-blade variable-pitch propellers. Tricycle landing gear, the two main units of which were retractable.
History: The Manx was ready for flight tests at the outbreak of the Second World War but, owing to pressure of work connected with the Halifax production programme, it was not until 1943 that the first flight was made. It had, however, begun taxiing trials at Radlett on 6 September 1942, on which date the nosewheel collapsed. In December of the same year, a propeller shaft and bearings were damaged when a ground crew member accidentally stepped into the rotating propeller and was killed. In March 1944, the original fixed landing gear was changed for a retractable type, and speed trials began in September 1945. The aircraft was eventually scrapped in 1952.

(UK)

HAWKER P.1052

First flight: 19 November 1948
Purpose: To investigate the behaviour of sweptback wings at low speeds, in terms of controllability and stability
Power plant: One Rolls-Royce Nene turbojet engine (5,000 lb; 2,268 kg st)
Wing span: 31 ft 6 in (9.60 m)
Length: 39 ft 7 in (12.07 m)
Height: 10 ft 6 in (3.20 m)
Wing area: 258 sq ft (23.97 m^2)
Weight loaded: 13,488 lb (6,118 kg)
Max level speed: about 564 knots (650 mph; 1,046 km/h)
Service ceiling: 45,500 ft (13,870 m)
Accommodation: Pilot only
Special design features: Same fuselage, tail unit (originally), landing gear and power plant as the Hawker Sea Hawk. Wings of symmetrical aerofoil section, with a sweepback of 35° at quarter chord and an aspect ratio of 3.84. Each wing in two parts, comprising a stub-wing and outer panel.

History: Built originally to Specification E.38/46, the P.1052 made use of many parts of the Hawker Sea Hawk. The spar-attachment frames in the centre fuselage were strengthened and the air intakes in the leading edges of the stub-wings were altered slightly because of the modified wing roots. In 1951, the first P.1052 was modified as a naval aircraft, with arrester-gear and other specialised equipment, and was used to provide experience in flying sweptwing aircraft from aircraft carriers. The same aircraft was later given a sweptback tailplane.

Meanwhile, in 1950, the second prototype P. 1052 had been converted into the P.1081 prototype fighter and first flew in its new form on 19 June 1950. It retained the front fuselage, wings and landing gear of the P.1052 but, to make provision for the ultimate use of afterburning, was given a new rear fuselage with straight-through jet-pipe. The P.1081 also had a new tail unit, with sweepback on all surfaces, and provision was made for four 20 mm cannon. In addition to development flying, the P.1081 was used to extend the P.1052's research programme of flight at high Mach numbers. It was destroyed in an accident on 3 April 1951.

(UK)

HAWKER SIDDELEY P.1127 KESTREL

First flight (tethered): 21 October 1960
Purpose: Experimental V/STOL tactical fighter
Power plant: One Bristol Siddeley Pegasus 5 vectored-thrust turbofan engine (15,200 lb; 6,895 kg st) with four rotating exhaust nozzles
Wing span: 22 ft 11 in (6.98 m)
Length overall: 42 ft 6 in (12.95 m)
Height: 10 ft 9 in (3.28 m)
Max T-O weight (VTOL, approx): 12,400 lb (5,625 kg)
Max level speed at S/L: Mach 0.87
Accommodation: Pilot only
Special design features: Cantilever shoulder-wings; anhedral 10° and sweepback at quarter-chord 32°. Anhedral tailplane. Retractable zero-track landing gear, with single nosewheel, twin main wheels and small outrigger wheels which retracted into wingtip fairings. Turbofan engine with four rotating exhaust nozzles.
History: Development of the Kestrel V/STOL tactical fighter was started in 1957 as a private venture. In 1959-60 the Ministry of Aviation ordered two prototypes and four development aircraft, which were known by the Hawker designation P.1127. The first prototype (XP831) began tethered hovering tests in October 1960; the first untethered hovering flight followed on 19 November, and conventional flight trials at RAE, Bedford, on 13 March 1961. Shortly afterwards it was joined by the second prototype. These tests culminated in the first transitions from vertical to horizontal flight and vice versa, by XP831, in September 1961. Subsequent flight trials included operation from the aircraft carrier *Ark Royal*, at sea, early in 1963. In the previous year, Britain, the United States and the Federal German Republic had announced a joint order for nine field evaluation Kestrels. The first evaluation aircraft flew for the first time on 7 March 1964 and introduced further design changes. The designation Kestrel F(GA) Mk 1 was given to this version in October 1964. Subsequent test flying led to an increase in tailplane span, and a small increase in the chord of the outer wing panels. All nine aircraft were delivered to a tripartite squadron based at RAF West Raynham, Norfolk, for evaluation flying between April and September 1965. Six Kestrels were shipped to the USA in early 1966, where they spent four months flying with the US Army, Navy and USAF. Following this, four were sent to Edwards AFB, California, and two were flown to NASA's Langley facility. The Kestrels remaining in the UK had, meanwhile, become associated with the flight development programme for the fully-developed Harrier.

(USA)

HILLER X-18

First flight (as conventional aeroplane): 24 November 1959
Purpose: Experimental tilt-wing convertiplane
Power plant: Two Allison T40-A-14 turboprop engines (each 5,850 eshp), driving six-blade Curtiss-Wright contra-rotating propellers, plus one Westinghouse J34 turbojet engine (3,400 lb; 1,542 kg st)
Wing span: 48 ft 0 in (14.63 m)
Length: 63 ft 0 in (19.20 m)
Height: 24 ft 7 in (7.49 m)
Weight loaded: 33,000 lb (14,969 kg)
Max level speed: 217 knots (250 mph; 402 km/h)
Max speed at which wing could be tilted: 155 knots (178 mph; 286 km/h)
Special design features: Fuselage of a Chase YC-122 transport aircraft. High-set wing with aspect ratio of 4.36. Incidence varied from 4° normal to 90° at maximum tilt. Vertical take-off setting 87°. Pitch control changed from tailplane and elevators in horizontal flight to jet-diverter in vertical flight, using a standard Westinghouse turbojet engine mounted in rear fuselage and provided with an extended tailpipe and jet diverter.
History: In February 1957 the US Air Force awarded Hiller an initial contract for the development of a twin-engined tilting-wing convertiplane to be designated X-18. Although it was the largest of the higher-speed VTOL aircraft built in the United States up to that time, and was unconventional by contemporary standards, its construction did not require the use of a large number of completely new and unconventional components. The X-18's wings, transmissions, engines and systems were essentially the same as those used in normal fixed-wing aircraft. The engines and their 16 ft propellers were obtained from the abandoned US Navy VTOL "tail-sitting" aircraft programme. The wing was designed to pivot through 90° for vertical take-off, so that the propellers worked in a similar way to the rotors of a helicopter. During cruising flight the X-18 flew like a conventional transport aircraft, but needed much less wing area than usual.

The X-18 flew for the first time, from Edwards AFB, as a conventional aircraft. In over 100 hours of subsequent ground and air tests, including 20 successful flights, it operated as a fixed-wing aircraft. Its wings were tilted progressively to an angle of attack of 50° before the programme was interrupted in 1961, so that the aircraft could enter a special powered rig for studies of downwash effects on ground environment during simulated hovering. Data provided by the X-18 was used in the development of the Chance Vought-Hiller-Ryan XC-142A tilt-wing aircraft.

(USA)

HUGHES XH-17

First flight: 23 October 1952
Purpose: Experimental heavy-lift helicopter designed to prove the pressure-jet propulsion system
Power plant: Two modified General Electric GE 5500 turbojet engines
Rotor diameter: 130 ft (37.62 m)
Height: over 30 ft (9.14 m)
Max T-O weight: 43,000 lb (19,960 kg)
Special design features and History: The XH-17 was an experimental heavy-lift helicopter which was built under contract to the USAF. Built primarily as a ground test model, the XH-17, after satisfactory tests of the jet-powered rotor mechanism, was converted into a flight test vehicle. It was powered by two specially-modified engines, developed and built by General Electric, which supplied gas pressure through ducts leading up the rotor shaft and out of the tips of the rotor blades. The XH-17 stood on four very high legs, which carried large wheels, and was thought of as the forerunner of powerful cargo-carrying helicopters that would lift and deliver equipment such as heavy artillery and bridge sections. It successfully completed its flight test programme between 1952 and 1953. The advantages accrued through the use of this system included a reduction in overall weight because of the elimination of the rotor transmission drive system. As follow-on to the XH-17, the XH-28 was designed, with a maximum weight of 105,000 lb. Only a mock-up was built, however, and the XH-28 programme was abandoned subsequently.

(USA)

HUGHES XV-9A

First flight: November 1964
Purpose: Research helicopter utilising a hot-cycle propulsion system
Power plant: Two General Electric YT64-GE-6 gas generators (each 2,850 shp)
Rotor diameter: 55 ft 0 in (16.76 m)
Length of fuselage: 45 ft 0 in (13.72 m)
Height to top of rotor head: 12 ft 0 in (3.66 m)
Design max T-O weight: 15,300 lb (6,940 kg)
Accommodation: Seating for two
Special design features: Three-blade rotor, with conventional pitch-change and flapping hinges but without drag hinges. Each blade was of constant 2 ft 7½ in (0.80 m) chord, attached to hub by two packs of retention straps. Gas duct of Rene 41 high-temperature steel passed between spars. Cooling air passed through leading- and trailing-edges of blade and exhausted at tips. Gas generator power plants were mounted in pods at ends of short high-set stub wings on each side of fuselage. Butterfly tail unit comprised two movable rudders. At low speeds and in VTOL flight the rudders were supplemented by hot gas jets expelled on each side of rear fuselage.
History: Under a US Army contract, Hughes designed, built and flew this comparatively large research helicopter utilising a hot-cycle propulsion system. The hot efflux from the turbines was ducted to nozzles at the blade-tips of the three-blade rotor, and to a yaw control jet at the tail. At each blade-tip, the gases were deflected through 90° by two sets of cascade vanes and accelerated to near sonic velocity.
As a first stage in the development programme, Hughes completed 60 hours of test running with a prototype hot-cycle rotor on a ground rig in 1962. The complete propulsion module (power plant, stub wing, rotor assembly) of the XV-9A was similarly tested for 15 hours in the Spring of 1964 before being installed on the fuselage. To cost as little as possible the aircraft itself was built of parts from other aircraft, including an OH-6A cockpit, two T64 engines lent by the US Navy, and an H-34 undercarriage. Flight tests began at Culver City on 5 November 1964 and the initial 15-hour flight test programme was completed on schedule on 5 February 1965. A further 20-hour test programme was then initiated on 30 April 1965 and was completed on 16 August 1965 after 19.1 flying hours. Although the XV-9A proved the concept of the hot cycle propulsion system, the aircraft itself was built only as a test-bed and proved to have poor handling characteristics, bad stability about all axes and poor control.

(Argentine Republic)

I.Ae.27 PULQUI (Arrow)

First flight: 9 August 1947
Purpose: Experimental fighter
Power plant: One Rolls-Royce Derwent 5 turbojet engine (3,500 lb; 1,588 kg st)
Wing span: 36 ft 11 in (11.25 m)
Length: 32 ft 8 in (9.96 m)
Height: 11 ft 1½ in (3.39 m)
Wing area: 212 sq ft (19.70 m^2)
Weight loaded: 7,936 lb (3,600 kg)
Max level speed: 459 knots (528 mph; 850 km/h)
Service ceiling: 50,850 ft (15,500 m)
Accommodation: Pilot only
Special design features: High-speed laminar-flow wing section. All-metal slotted flaps, with differential operation for use as ailerons when landing. Nose air duct, which bifurcated to by-pass cockpit. Fuel tanks in wings. Four 20 mm cannon in nose.

History: The Pulqui was the first jet-propelled aircraft to be designed, built and flown in Latin America. Designed by M. Emilio Dewoitine, assisted by Commodore Juan San Martin, it was a simple aircraft with straight wings, an elliptical-section fuselage with the jet orifice at the rear, a conventional tail and a tricycle landing gear. Had the design developed further, it was proposed to fit the Pulqui with jettisonable auxiliary fuel tanks, to increase its limited range, and to arm it with rockets or bombs for an alternative attack role.

(Argentine Republic) I.A.38

First flight: 9 December 1960
Purpose: Experimental tail-less cargo transport aircraft
Power plant: Four I.A. 16El Gaucho radial piston engines (each 450 hp) driving pusher propellers
Wing span: 105 ft 0 in (32 m)
Length: 44 ft 3½ in (13.5 m)
Height: 15 ft 1 in (4.6 m)
Wing area: 1,431.5 sq ft (133 m^2)
Weight loaded: 35,274 lb (16,000 kg)
Max level speed (estimated): 135 knots (156 mph; 252 km/h)
Service ceiling (estimated): 14,775 ft (4,500 m)
Range (estimated): 674 nm (776 miles; 1,250 km)
Accommodation: Crew of two
Special design features: Swept wing. Special wing profile, with 18% thickness/chord ratio at root and 10.4% at tip. Aspect ratio 7.7. Chord 21 ft (6.4 m) at root, 5 ft 6 in (1.6 m) at tip. Sweepback at leading-edge 36.5°. Fuselage incorporated in and below wing. Fins and balanced rudders near wingtips. Retractable landing gear with one wheel at nose and four main wheels. Large compartment capable of accommodating 6 tons of cargo within and below wing.
History: Designed and built under the direction of Dr Reimar Horten, the I.A.38 was a large tail-less monoplane based on the German wartime Horten Ho VIII project. The prototype was completed in 1959 but flight trials were delayed because of troubles with engine cooling. The first flight took place in late 1960 but the project was suspended soon afterwards.

(USSR) KAMOV Ka-22 VINTOKRYL

First public appearance: 9 July 1961
Purpose: Heavy-lift convertiplane
Power plant: Two Soloviev TV-2 shaft-turbine engines (each 5,622 ehp)
Wing span: approx 92 ft 0 in (28.04 m)
Accommodation: Two pilots; may have been designed to carry between 80 and 100 passengers
Special design features: Fuselage was approximately the same size as that of the Antonov An-12, with upswept tail which incorporated a loading ramp for vehicles or freight. Entire trailing-edge of high-set tapered cantilever wing was made up of ailerons and flaps. Non-retractable tricycle landing gear with twin wheels on each unit. Engines mounted in wingtip pods, with tailpipes which could be tilted downward for additional lift in vertical flight. Engines drove four-blade rotors mounted above the engines during take-off, landing, hovering and low-speed flight; power was transferred to the forward-facing four-blade propellers during cruising flight, when the rotors autorotated.
History: One of the surprises of the 1961 Soviet Aviation Day display was this large twin-engined convertiplane, which the commentator described as the most powerful vertical take-off machine in the world. On 7 October 1961, the Vintokryl set up a Class E.2 convertiplane speed record of 192.3 knots (221.4 mph; 356.3 km/h) over a 15/25 km course. On 24 November 1961, it lifted a payload of 33,069 lb (15,000 kg) to a record height of 8,491 ft (2,588 m), qualifying also for records with a payload of 1,000, 2,000, 5,000, and 10,000 kg. On the same day, it lifted to 6,562 ft (2,000 m) a payload of 36,343 lb (16,485 kg). The pilots on each occasion were D. Efremov and G. Gromov.

(France)

LEDUC 0.10

First flight (powered): 21 April 1949
Purpose: Experimental aircraft powered by a ramjet engine
Photo: Leduc 0.10 on its launching aircraft
Power plant: One Leduc ramjet engine (4,400 lb; 1,995 kg st)
Wing span: 34 ft 6¼ in (10.52 m)
Length: 33 ft 7½ in (10.25 m)
Wing area: 172.22 sq ft (16 m²)
Weight loaded: 6,615 lb (3,000 kg)
Max level speed: Mach 0.84
Accommodation: Crew of two
Special design features: Tubular double-skinned fuselage. Inner fuselage shell contained pilot's cockpit; outer shell formed annular duct for ramjet. Injector pumps and generator powered by an auxiliary gas turbine situated behind pilot's cockpit, aft of which were five cylindrical ducts of increasing size. Leading-edge of each duct ringed with fuel injectors, with 500 burners in all. Nose-cone with perspex surround for pilot's forward vision. Circular windows in sides of outer duct for lateral vision. Forward portion of fuselage detachable in an emergency; parachute for lowering the capsule contained in upper fuselage. Tapered wings and tail surfaces.

History: René Leduc concentrated for many years on development of an athodyd or ramjet power plant for aircraft. As far back as 1935 he produced a small unit which developed a thrust of 8.8 lb (4 kg) at 984 ft/sec (300 m/sec). He subsequently went on to build the 0.10 aircraft, which was first released as a glider from the back of an SE 161 Languedoc carrier aircraft in October 1947. Because the ramjet power plant could not operate until sufficient air velocity was introduced into the tube, the powered version of the Leduc 0.10 was also taken into the air on an SE 161 carrier aircraft and, in April 1949, the first powered flight was made over Toulouse. The 0.10 was released at about 174 knots (200 mph; 320 km/h), at which speed the fuel could be turned on and ignited. The flight lasted 12 minutes, and the 0.10 attained a speed of about 366 knots (422 mph; 680 km/h) on half power. During a later flight the aircraft flew at about 434 knots (500 mph; 805 km/h), still on half power, at an altitude of 36,100 ft (11,000 m). Two more examples were built. The first was identical to the original 0.10; the second, the 0.16, was fitted initially with wingtip-mounted Turboméca turbojet engines. This first flew on 8 February 1951, but the turbojet engines were soon removed and balances were installed in their place.

(France)

LEDUC 0.21 and 0.22

First flight (0.21): 16 May 1953
Purpose: Experimental ramjet monoplanes
Photo and Data: 0.21
Power plant: Leduc "athodyd" (or ramjet) fuselage duct developing 14,330 lb (6,500 kg) st at 621 mph (1,000 km/h)
Wing span: 38 ft 0¾ in (11.6 m)
Length: 41 ft 0 in (12.5 m)
Height: 9 ft 0¼ in (2.75 m)
Weight loaded: 13,227 lb (6,000 kg)
Max level speed: Mach 0.87
Service ceiling: 65,625 ft (20,000 m)
Endurance: 15 min
Accommodation: Pilot only
Special design features: Bi-convex wing section, tapered from 10% thickness/chord at root to 8% at tip. Aspect ratio 5.5. Leading-edge sweepback 14°. Dihedral 3°. Wingtip nacelles for balancer wheels of landing gear. Tubular open-ended fuselage duct of double-skinned structure mounted on the rear part of cockpit nacelle by supports. Retractable tandem main wheels and retractable tailwheel. Forward portion of cabin nacelle detachable in emergency, and lowered to ground by three parachutes.

Engine operation: Air entered annular opening surrounding pilot's cockpit, and passed to centre of fuselage where it entered a series of six internal cylinder ducts of increasing size, the leading-edge of each duct being ringed with fuel injectors. Gases produced by ignition of the resultant mixture were ejected from the rear end of the fuselage as a high-velocity jet. Engine was lit by the exhaust from a Turboméca Artousté turbine mounted at the aft end of the central nacelle.

History: The Leduc 0.21 was about one-third larger than the earlier 0.10. Two prototypes were built, and had to be launched from a carrier aircraft because this type of power plant could not operate before sufficient air velocity was introduced into the tube. The 0.21s proved completely successful and were used to flight test components and equipment for the 0.22 Mach 2 interceptor. This latter aircraft was a development of the 0.21, both types being developed under an order from the French Air Ministry. Of more cylindrical appearance, the first of two 0.22s flew on 26 December 1956. The 0.22 could take-off under its own power, with the aid of an Atar D.3 turbojet engine, and by March 1957 it had completed over 30 test flights on the power of the Atar alone. However, official support for the aircraft was withdrawn under the 1957 cut-backs in military contracts, and development ceased.

(USA)

LOCKHEED XFV-1

First flight: March 1954
Purpose: Experimental VTOL "tail-sitting" fighter
Photo: XFV-1 on handling trolley
Power plant: One Allison T40-A-6 turboprop engine (5,850 ehp)
Wing span: about 24 ft 0 in (7.32 m)
Length: about 28 ft 0 in (8.53 m)
Height (tail, aircraft in horizontal position): about 11 ft 0 in (3.35 m)
Max level speed: about 434 knots (500 mph; 805 km/h)
Special design features: Straight wings of long chord at the roots. Cruciform tail unit, with a sprung swivelling castor wheel at extremity of each surface for vertical take-off and landing. Circular-section fuselage, with air intakes on sides and underneath. High cockpit enclosure and gimbal-mounted pilot's seat, able to tilt at 45° with aircraft in vertical position. Two 16 ft (4.88 m) diameter Curtiss-Wright Turboelectric co-axial contra-rotating propellers. Conventional, but high, landing gear for horizontal take-offs and landings during flight trials. Special ground-handling trolley used for translating the aircraft between horizontal and vertical positions. Mobile cockpit access ladder for pilot.
History: The US Navy held, in 1950, a design competition for a fighter that could take off and land vertically, and carry out normal aircraft manoeuvres in horizontal flight. Lockheed and Convair were awarded contracts to build prototypes (the Convair aircraft, the XFY-1, can also be found in this book), and the Lockheed XFV-1 was the first to fly. Although trials were carried out in horizontal flight, the XFV-1 was never flown vertically. The US Navy cancelled its contract with Lockheed and further development was abandoned.

(USA)

LOCKHEED XH-51A (MODEL 186)

First flight: 2 November 1962
Purpose: Research helicopter
Data: XH-51A
Photo: XH-51A Compound
Power plant: One Pratt and Whitney (Canada) T74 (PT6B) shaft-turbine engine (500 shp)
Main rotor diameter: 35 ft 0 in (10.67 m)
Length overall: 40 ft 9 in (12.4 m)
Height: 8 ft 2½ in (2.50 m)
Max normal T-O weight: 4,100 lb (1,860 kg)
Max level speed at S/L: 151 knots (174 mph; 280 km/h)
Hovering ceiling in ground effect: 16,000 ft (4,876 m)
Range with max fuel: 225 nm (260 miles: 418 km)
Accommodation: Crew of two
Special design features: Four-blade rigid main rotor. Features of the design included cantilevered blades, with freedom in the feathering axis only; a mechanical stabilising gyro located in series between the blades and pilot's controls; and a control system utilising a spring cartridge between the pilot's stick and the control gyro to provide the required mechanical motion and force characteristics. Retractable landing-gear skids. Two flush NASA airscoops for the engine in top of fuselage aft of main rotor mast fairing. Fixed horizontal stabiliser at tail to compensate for fuselage aerodynamic pitching moment.
History: In February 1962, the US Army and Navy awarded Lockheed-California Company a contract to build two high-performance helicopters. These had rigid rotors of the kind pioneered on the Lockheed CL-475 flying testbed helicopter. Advantages claimed for the rigid rotor system included inherent stability without artificial stabilisation, high manoeuvrability, extremely wide usable CG range, low vibration and ease of control. With the XH-51A Lockheed achieved an aerodynamic drag value as low as that of a flat plate only 8 sq ft (0.74 m^2) in area. External skins were flush-riveted, and the fuselage was flush-sealed extensively. To reduce further the fuselage drag at high speed, the rotor plane was set at a forward tilt of 6° relative to the fuselage datum line. Three versions were eventually tested, as the XH-51A, XH-51A Compound and XH-51N. Of these, the Compound, which had a P & W J60-P-2 auxiliary turbojet engine (2,600 lb; 1,180 kg st) and short-span wings, established an unofficial record for rotating-wing aircraft of 263 knots (302.6 mph; 487 km/h) in June 1967. The XH-51N was similar to the "A" but had a three-blade rotor and could seat five.

(USA)

LOCKHEED XV-4A HUMMINGBIRD

First flight (conventional): 7 July 1962
Purpose: Two-seat VTOL research aircraft
Power plant: Two Pratt and Whitney JT12A-3 (mod) turbojet engines (each 3,300 lb; 1,497 kg st)
Wing span: 25 ft 11½ in (7.91 m)
Length overall: 33 ft 11 in (10.34 m)
Height: 11 ft 9 in (3.58 m)
Max vertical T-O and landing weight: 7,200 lb (3,265 kg)
Max level speed at S/L (estimated): 452 knots (520 mph; 835 km/h)
Special design features: Short wings; aspect ratio 6, chord 6 ft (1.83 m) at root and 2 ft 4 in (0.71 m) at tip, no dihedral. Ailerons and single-slotted trailing-edge flaps along entire trailing-edge of wings. Wingtip "salmons". Engine bleed air ejected over upper surface of tailplane and elevators for BLC.
History: In June and September 1961, the US Army Transportation Research Command awarded Lockheed-Georgia Company contracts for two small research aircraft, to prove the practicability of a new VTOL technique known as augmented-jet ejector lift. In the Hummingbird, the engines provided thrust for both VTOL and forward flight, by means of a system of diverter valves in the tailpipes. For take-off and landing, the valves diverted the entire jet efflux from both engines through 180° into a pair of ejector ducts. The efflux was discharged downwards from the ducts, through 20 transverse rows of multiple nozzles, into the two ejector chambers. In the top and bottom of the fuselage were bomb-bay type doors, which were opened during VTOL flight. When the high-velocity exhaust gases were ejected downward through the ejector chambers, they drew with them a quantity of free air through the open doors. This air was accelerated by the hot gases issuing from the ejector nozzles and boosted the vertical lift by about 40%. The ejector chambers were angled rearward at 12° to the vertical. After vertical take-off, the aircraft's nose was tilted down about 10° and it began to move forward. At about 80 knots (92 mph; 148 km/h) and 126 knots (145 mph; 232 km/h) respectively, one and then two engines were switched to forward thrust and the nose was raised. During vertical and low-speed flight, stability and control were achieved by reaction jet nozzles at the nose, tail and wingtips. Following static tests, the XV-4A made its initial tethered hovering flight on 30 November 1962. Free hover flight tests began in the Summer of 1963, and on 8 November 1963 the first transition from vertical to horizontal flight was made successfully. However, the aircraft crashed on 10 June 1964, killing the pilot. The second prototype was converted for direct jet-lift research, as the XV-4B Hummingbird II, but crashed in 1969.

MARTIN MARIETTA X-24A (SV-5P PILOT) (USA)

First flight (powered): 19 March 1970
Purpose: Lifting-body research aircraft designed for the lower speed scale
Power plant: One Thiokol XLR11 four-chamber regeneratively-cooled turbo-rocket engine (8,000 lb; 3,625 kg st); two Bell LLRV landing rockets (each 500 lb; 225 kg st)
Width overall: 13 ft 8 in (4.17 m)
Length overall: 24 ft 6 in (7.47 m)
Height: 10 ft 4 in (3.15 m)
Planform area: 162 sq ft (15.05 m²)
Max launching weight: 10,700 lb (4,853 kg)
Accommodation: Pilot only
Special design features and History: Martin Marietta has been engaged in lifting-body research and development since 1959. The aim was to pioneer the development of manoeuvring manned re-entry vehicles able to perform as spacecraft in orbit, fly in Earth's atmosphere like aircraft, and land at conventional airports. The small unmanned X-23A first proved the aerodynamic characteristics of the basic design. The SV-5P Pilot (Piloted Law-speed Test aircraft) was ordered by the USAF in May 1966. Known officially as the X-24A, it was of triangular planform and "bulbous wedge shape", with a flat bottom, rounded top and three vertical fins. Controls consisted of two upper flaps (aileron and elevator) and two lower flaps at the extreme rear of the body, and a pair of split rudders on each outer tail-fin. Construction was conventional, using aluminium alloys. The X-24A was delivered to the USAF on 11 July 1967, and was sent to Edwards AFB for flight testing after the initial acceptance tests. During 1969 the X-24A completed successfully nine unpowered flights, and the first powered flight was then made by Maj Jerauld Gentry, USAF. Total flight time on this occasion was 7 min 15 sec, with a 2 min 40 sec burn of the Thiokol rocket engine. An adapter enabled the X-24A to be carried into the air under a B-52 "mother-plane". After release at about 45,000 ft (13,700 m) and speed of about Mach 0.6, the XLR11 engine was ignited to boost the X-24A to maximum speed and height. It was then manoeuvred to a landing on Rogers Dry Lake, at Edwards AFB. A total of 28 flights were made with the X-24A during this programme; several flights were supersonic, and a maximum speed of Mach 1.62 and altitude of 71,407 ft (21,765 m) were attained in 1971. It was announced by NASA in that year that the X-24A was to be stripped down to its basic structure and rebuilt as the X-24B.

(USA)

MARTIN MARIETTA X-24B

First flight: 1 August 1973
Purpose: Lifting-body research aircraft
Power plant: One Thiokol XLR11 four-chamber regeneratively-cooled turbo-rocket engine (8,000 lb; 3,625 kg st); two Bell LLRV optional landing rockets (each 400 lb; 181 kg st)
Width overall: 19 ft 2 in (5.84 m)
Length overall: 37 ft 6 in (11.43 m)
Height: 10 ft 4 in (3.15 m)
Double-delta planform area: 330 sq ft (30.66 m²)
Max T-O weight: 13,000 lb (5,896 kg)
Max level speed: 1,005 knots (1,158 mph; 1,863 km/h)
Service ceiling: 74,132 ft (22,595 m)
Accommodation: Pilot only
Special design features: Fuselage structure of light alloy, primarily 2024 aluminium. Triangular cross-section, with flat bottom and rounded top. Triple-finned tail unit; fixed centre fin; outer fins each carried a pair of split rudders. Upper and lower flaps at extreme tail, between fins, served as elevators and for pitch trim. Ailerons outboard of the fins. Manually-retracted tricycle landing gear.

History: It was announced by NASA on 29 July 1971 that the X-24A was to be stripped down to its basic structure and rebuilt as the X-24B, with completely new external lines. This work started in January 1972 and was completed in October of the same year. Following its first unpowered flight in August 1973, the new research aircraft made five more unpowered tests and 13 highly successful powered flights by the Spring of 1975. Its final powered flight was made on 23 September 1975, at NASA's Flight Research Center, making it the last known rocket-powered aircraft to fly and thus ending an era in experimental flight research. Six more gliding flights were planned at that time.

(USA)

McDONNELL XF-85 GOBLIN

First flight: 23 August 1948
Purpose: Parasite jet-propelled interceptor fighter
Photo: XF-85 hooked-on and then released from the B-29 "mother-plane"
Power plant: One Westinghouse 24C (J34-WE-22) turbojet engine (3,000 lb; 1,361 kg st)
Wing span: 21 ft 2¾ in (6.47 m)
Length: 14 ft 10½ in (4.53 m)
Height: 8 ft 3¼ in (2.52 m)
Weight loaded: about 4,835 lb (2,193 kg)
Max level speed: about 452 knots (520 mph; 837 km/h)
Accommodation: Pilot only
Special design features: Extremely stubby monocoque fuselage, with retractable "sky-hook" ahead of cockpit. Folding wings, with 34° sweepback and 4° anhedral. To avoid folding for stowage aboard a bomber, six tail surfaces were spaced round the rear fuselage; two wingtip fins were fitted later (see below). No landing gear, although skids were fitted for emergency use during flight testing. Air intake in nose
History: This parasite fighter was originally developed as part of the defence system of the Consolidated-Vultee B-36 six-engined long-range bomber. The XF-85 was designed to be carried in the forward bomb-bay of the B-36, and to be launched and picked up by a release and hook-on "trapeze" technique. It was intended that while the aircraft was stowed in the bomb-bay of the mother-plane, it could be refuelled and receive radio, radar and mechanical maintenance. A contract covering the initial development phase was awarded to McDonnell by the Air Force on 9 October 1945, and in 1947 Lockheed F-80s were used to simulate hook-ons by XF-85s when completed. Before free flights were made the XF-85 was taken aloft by its mother aircraft and tested in a captive mode. The first free flight test was made from a specially-equipped Boeing B-29 over Muroc on 23 August 1948. However, the first attempt to hook-on at 25,000 ft was hampered by rough air. The B-29 trapeze fouled the canopy of the XF-85, smashing the Plexiglas, and hit the pilot on the head, knocking off his helmet and oxygen mask. The pilot put the oxygen hose in his mouth and made a successful emergency landing on the skids at nearly 170 mph (275 km/h). Because the XF-85 proved unstable in flight, the effectiveness of the original tailplane arrangement was increased by the addition of wing-mounted fins. Three successful hook-ons were achieved later out of seven flights; the first hook-on on 14 October 1948. But the concept of using the XF-85 as part of the defences of the B-36 was abandoned. Only two aircraft were built.

(USA)

McDONNELL XF-88B

First flight: 14 April 1953
Purpose: Research into supersonic propeller design
Photo: XF-88B
Power plant: Two Westinghouse J34 turbojet engines (each 3,000 lb; 1,360 kg st) and one Allison XT38 turboprop engine
Wing span: 39 ft 8 in (12.09 m)
Length (XF-88): 54 ft 1½ in (16.50 m)
Height (XF-88): 17 ft 3¼ in (5.26 m)
Weight loaded (designed, XF-88): about 15,000 lb (6,800 kg)
Accommodation: Pilot only
Special design features: Very thin laminar-flow wings, swept at 35°. Two turbojet engines with air intakes in wing roots, exhausting behind wings and below tail. Nose-mounted turboprop engine, fitted during tests with different propellers, including Curtiss Electric Hamilton Standard and Aeroproducts supersonic types of 4 ft, 7 ft and 10 ft diameter, and of two-, three- and four-blade configuration. Engine gearboxes of three different ratios to match engine/propeller speeds.
History: Detail design of an experimental long-range penetration fighter was started by McDonnell in June 1946 and the first example, the XF-88, made its maiden flight on 20 October 1948. It was followed in 1950 by the XF-88A, a version that was similar to the XF-88 but had its turbojet engines fitted with short afterburners of McDonnell design, giving 3,600 lb (1,633 kg) st. Because of changes in USAF requirements and defence cutbacks, the further development of these aircraft was cancelled in August 1950. After a period of more than a year, the XF-88A was revived and, with modifications, was ordered into production as the McDonnell Voodoo long-range escort fighter, under the designation F-101. Subsequently, one of the two prototypes of the XF-88 was fitted with a turboprop engine, in addition to the turbojet engines, and began testing supersonic propellers. In this configuration the aircraft was redesignated XF-88B. Normal power was provided by the turbojet engines, the turboprop being used only for propeller research.

(USA)

McDONNELL XV-1

First flight (first translation from vertical to horizontal flight): 29 April 1955
Purpose: Experimental convertiplane
Power plant: One Continental R-975-19 piston engine (550 hp) and McDonnell rotor-tip pressure-jets
Rotor diameter: approx 26 ft 0 in (7.92 m)
Length: approx 30 ft 0 in (9.14 m)
Height: approx 10 ft 8 in (3.25 m)
Accommodation: Pilot and provision for three passengers, or two stretchers and medical attendant instead of passengers

Special design features and History: The XV-1 experimental convertiplane was a joint development of the Wright Air Development Center, USAF, the Transportation Corps of the US Army and the McDonnell Aircraft Corporation. It used a jet-driven rotor for vertical flight, and wings and a normally-driven propeller for forward flight, during which the rotor autorotated at its lowest drag configuration. Each of the three rotor blades was powered by a McDonnell pressure-jet unit located at the blade tip. A Continental engine in the rear fuselage drove two compressers to supply air to the jet units during vertical flight, and drove the 6 ft 5 in (1.96 m) diameter Met-L-Prop two-blade propeller during forward flight. Fuel was fed to the rotor-tip burners through a rotary fuel governor driven from the rotor hub accessory drive. There were also two small anti-torque tail rotors. The tail assembly consisted of two vertical fins and rudders attached to the extremities of the twin tailbooms. A freely-floating horizontal tail surface was mounted between the booms.

Design of the XV-1 was initiated in mid-1951, the first prototype being rolled out in January 1954. Following ground tests, which were carried out between 23 January and 10 February 1954, the first take-off was made on 11 February. The second prototype made its first flight on 14 July 1954 but was damaged on 10 December during power-off autorotation landing trials. The first full conversion was accomplished successfully on 29 April 1955.

During evaluation trials in August 1956, one of the two prototypes was claimed to have set an unofficial helicopter speed record of 174 knots (200 mph; 322 km/h). Development of the XV-1 was discontinued in 1957.

(USSR)

MIKOYAN E-166

First flight: Unknown
Purpose: Experimental aircraft, probably built as a testbed for aerodynamic and power plant research
Power plant: One TDR Mk P.166 turbojet engine (22,046 lb; 10,000 kg st)
Wing span (estimated): over 26 ft 0 in (8.10 m)
Length: over 55 ft 0 in (16.76 m)
Max level speed: Mach 2.52
Accommodation: Pilot only
Special design features and History: Until July 1967 little was known of this aircraft, except what was disclosed by the Soviet authorities when submitting claims for international records established by it. When displayed in the static park at the Domodedovo air display in that month, it was seen to be similar in configuration to the MiG-21, but scaled up. The mid-set wings and all-moving horizontal tail surfaces were similar in planform to those of the MiG-21. The small one-piece curved canopy was flush with the very large dorsal "spine". Other features included a large conical centrebody in the air intake, and a shallow full-chord fence under each wing at mid-span, supplemented by a small fence above each wing near the tip. A flat aerofoil-shape projection on each side of the front fuselage was probably intended to divert the boundary-layer airflow locally, forward of the wing.

First pilot to establish an international record in the E-166 had been A. Fedotov, who set up a speed record of 1,295.6 knots (1,491.9 mph; 2,401 km/h) over a 100 km closed circuit on 7 October 1961. The E-166 subsequently set up more officially-recognised records. On 7 July 1962, it averaged 1,446.73 knots (1,665.93 mph; 2,681 km/h) over a 15/25 km course at Podmoskovnoe aerodrome, to establish an absolute speed record. During one run, it was stated that a speed of over 1,620 knots (1,865 mph; 3,000 km/h) was achieved. On 11 September the E-166 set up a sustained altitude record of 74,376 ft (22,670 m) over a 15/25 km course at Jukovski-Petrovskoi, during which a speed of 1,349 knots (1,553 mph; 2,500 km/h) was maintained.

(USSR)

MIKOYAN "FAITHLESS"

First flight: Unknown
Purpose: Experimental STOL fighter
Photo: "Faithless" with lift-jet doors open
Power plant: One large afterburning turbojet engine, with variable-area nozzle. Two lift-jet engines mounted in tandem in the centre fuselage, between the air intake ducts for the propulsion engine.
Wing span (estimated): 30 ft 0 in (9.15 m)
Length overall (estimated): 60 ft 6 in (18.50 m)
Accommodation: Pilot only
Special design features and History: Second of the Mikoyan STOL fighters demonstrated at Domodedovo on 9 July 1967 was a completely new design. Although it followed the delta-wing/swept-tail formula of the MiG-21, it had larger overall dimensions. The fuselage had an ogival nosecone and was flattened on each side forward of the engine air intakes. These were of the semi-circular type, with half-cone centre-bodies; and there was an additional small inlet door in each intake duct above the wing leading-edge, to provide extra air for the engine at low forward speeds. The wings were mid-set, with the all-moving horizontal tail surfaces positioned a little lower on the rear fuselage. Large area-increasing flaps extended from the inboard end of the inset aileron to the fuselage on each side. No wing fences were fitted. A cruciform brake-chute was housed in an acorn fairing at the base of the rudder. A rearward-hinged trap-type intake box was fitted above the lift-jet engines, forming the top surface of the fuselage when closed. There were four rows of longitudinal slots in this upper door and a panel of transverse louvres under the belly of the aircraft, beneath the lift-jets. No conventional aircraft of this basic design, without a lift-jet installation, has been demonstrated in public. However, the fuselage and tail-unit of "Faithless" were similar in size and general outline to those of the prototype MiG-23 ("Flogger") variable-geometry fighter.

(USSR)

MIKOYAN MiG-21 ANALOGUE

First flight: Unknown
Purpose: Aerodynamic flight testing and development of a scaled-down replica of the "ogee" delta wing for the Tu-144 supersonic airliner.
Photo: Analogue escorting a prototype Tu-144
Power plant (MiG-21PF): One RD-11 turbojet engine (13, 120 lb; 5,950 kg st with afterburning)
Accommodation: Pilot only

Special design features and History: Based on a standard MiG-21PF airframe, this aircraft was fitted with a scaled-down replica of the Tu-144's wings for testing and development before the Tu-144 prototype was completed. It had no horizontal tail surfaces. As a result of its several dozen research flights, modifications were made to the full-size wings. Only one Analogue was built.

(USSR)

MIKOYAN MiG-21 "FISHBED-G"

First flight: Unknown
Purpose: Experimental STOL fighter
Power plant: One RD-11 turbojet engine (13,120 lb; 5,950 kg st) with afterburning. Two jet-lift engines mounted in tandem in the centre fuselage
Wing span: 23 ft 5½ in (7.15 m)
Length: approx 55 ft 8 in (16.97 m)
Accommodation: Pilot only
Special design features and History: Three different jet-lift STOL fighters were demonstrated during the air display at Domodedovo on 9 July 1967, which indicated the importance attached by the Soviet Air Force to this technique for increasing the versatility of its tactical units. One of the three aircraft was a fairly simple adaptation of a standard MiG-21PF, with an additional fuselage section some 4 ft (1.22m) long inserted aft of the cockpit to house two vertically-mounted lift-jet engines. The air supply for these was obtained by raising a rearward-hinged door panel, which formed the top skin of the fuselage when closed. Longitudinal slots and transverse louvres were the same as on "Faithless" (which see). The landing gear had a wider track than that of the standard MiG-21, and the main wheels appeared to be non-retractable, which implied that this aircraft was produced primarily as a lift-jet test vehicle.

(UK)

MILES M-52

First flight: Project abandoned before first flight
Purpose: Supersonic flight research
Photo: Model of the
Power plant: One Power Jets (Research and Development) Ltd W.2/700 turbojet engine with augmentor and afterburner (2,000 lb; 907 kg st at S/L and 4,100 lb; 1,860 kg st at maximum speed)
Wing span: 26 ft 10½ in (8.19 m)
Length: 33 ft 6¼ in (10.22 m)
Wing area: 108 sq ft (10.03 m²)
Max weight: 8,200 lb (3,719 kg)
Designed max speed: 868 knots (1,000 mph; 1,609 km/h) at 36,000 ft (10,975 m), after diving from 50,000 ft (15.240 m)
Accommodation: Pilot only
Special design features: Very thin bi-convex section wings, with knife-sharp leading- and trailing-edges. All-moving tailplane with the horizontal surfaces arranged to move longitudinally. Annular air intake at nose. Jettisonable pilot's cabin which, following ejection from the aircraft, would be lowered by parachute to an altitude where the pilot could bale out and descend with his own parachute.
History: The M-52 was specially designed for experimental flying at 1,000 mph at 36,000 ft, to which height it was expected to climb in 1½ minutes. Design work began in 1943, under an official contract, to Specification E.24/43; by February 1946, 90% of the detail design had been completed. For economic reasons, however, and because of a mistaken belief that sweptback wings were essential for supersonic flight, the contract was cancelled. The first flight had been intended for 1946, and test models built and flown by Vickers in 1947-48 proved that the aircraft would have performed its designed task. The wings and tailplane for the M-52 had previously been flight tested on a Miles Falcon-Six.

(USA) MISSISSIPPI STATE UNIVERSITY MARVEL AND MARVELETTE

First flight (Marvelette): March 1962
Purpose: Flying testbed research aircraft
Photo: XV-11A Marvel
Data: Marvelette
Power plant: One Continental C90-12 four-cylinder piston engine (95 hp)
Wing span: 26 ft 2 in (7.97 m)
Length overall: 28 ft 5 in (8.66 m)
Height: 7ft 11 in (2.41 m)
Propeller duct diameter, internal: 5 ft 6 in (1.68 m)
Max payload: 250 lb (113 kg)
Max T-O weight: 2,000 lb (907 kg)
Max level speed at S/L: 109 knots (125 mph; 201 km/h)
Accommodation: Seating for one or two
Special design features: Special glass-fibre camber-changing wing. Aspect ratio 6.5. Thickness/chord ratio 15%. Dihedral 2°. Distributed BLC suction through perforations. Conventional ailerons. Electrically-actuated camber-changing removed need for conventional flaps. Propeller duct replaced conventional tail unit. Elevator and rudder functions performed by movable segments in trailing-edge of duct. Trim-tab on elevator segment. Non-retractable landing gear. Engine drove an aluminium three-blade electrically-actuated variable-pitch propeller of MSU manufacture. Equipment included instruments to measure pressure inside wing and boundary layer control (BLC) flow.
History: Intended as a flying testbed for features of the Marvel, the Marvelette (US Army designation XAZ-1) was fitted with the latter's BLC wings and ducted propeller. The Marvelette logged 16 flights by March 1964. Numerous modifications were made and tested both in flight and on the ground, including installation of glass-fibre main landing gear. Results of this development were applied to the design of the Marvel. Preliminary and aerodynamic design of both aircraft was carried out at the MSU, where the fuselage of the Marvelette was also built. The wings, propeller, shroud-duct assembly and shaft were manufactured by Parsons Corporation of Traverse City, Michigan. The Marvel (XV-11A) was a STOL research aircraft which was put under development for the US Army. It had over one million tiny holes for BLC drilled in its wings and part of the fuselage. Power was provided by a 250 hp Allison T63 shaft-turbine engine. The prototype Marvel was completed in the Summer of 1966, and it began an extensive flight test programme at the University, under the sponsorship of the US Army Aviation Material Laboratories. Testing to obtain performance and stability/control data was completed during 1969. A programme of acoustic research was scheduled to start at a later date.

(USSR)

MYASISHCHEV M-52 "BOUNDER"

First flight: about 1958
Purpose: Experimental bomber
Power plant: Four D-15 turbojet engines (each 28,660 lb; 13,000 kg st), the inner two having afterburners
Wing span (estimated): 90 ft 0 in (27.43 m)
Length overall (estimated): 180 ft 0 in (54.86 m)
Max level speed (estimated): 1,079 knots (1,243 mph; 2,000 km/h)
Special design features: Shoulder-mounted delta wings, with slightly reduced sweepback on the outer panels. Forward-swept wingtips with one turbojet engine on end of each. Tandem eight-wheel main landing gear bogies and wingtip balancer wheels. Large-area vertical tail and swept horizontal tail surfaces, all fitted with anti-flutter weights.
History: Statements made by the commentator during the 1961 Aviation Day flypast indicated that this four-jet aircraft was designed by the Myasishchev team. The aircraft was first known to Western air forces some three years earlier and was described by General Thomas D. White, then Chief of Staff of the USAF, as being "probably configured for supersonic flight and in the 300,000 lb (136,000 kg) class". All four turbojet engines were believed to have been in underwing pods originally, the outer engines being transferred later to the wingtips. Only one example of the M-52 (known to NATO as "Bounder") was shown at Tushino, although at least one other was built. However, the M-52 proved to have a limited range, and the prototypes are thought to have been adapted for experimental work.

(USA/Canada) NASA/DITC AUGMENTOR WING JET STOL RESEARCH AIRCRAFT

First flight: 1 May 1972
Purpose: Augmentor wing research
Data: Modified C-8A Buffalo
Power plant: Two Rolls-Royce Spey Mk 801SF turbofan engines (9,000 lb; 4,082 kg st), each modified to include a new by-pass duct that collects the fan air and directs it to two 13 in (0.33 m) diameter offtake ducts on top of the engine. Vectored nozzle assembly on each engine in place of the normal tailpipe.
Wing span: 78 ft 9 in (24.00 m)
Total wing area: 865 sq ft (80.36m²)
Special design features and History: In a co-operative venture between NASA and the Canadian Department of Industry, Trade and Commerce (DITC), a de Havilland Canada C-8A Buffalo transport aircraft has been modified extensively to serve as an augmentor wing research aircraft. It was chosen for modification to test the concept because its high wing and T-tail made it very suitable for modification into a powered-lift jet STOL transport; its wing planform is basically similar to that of a wind-tunnel model previously tested. Major modifications and alterations to the aircraft included a reduction of wing span; replacement of all the original wing structure aft of the rear spar, by installation of an augmentor flap system, including augmentor chokes, installation of drooped ailerons with boundary layer control (BLC), and repositioning and redesign of spoilers; installation of fixed full-span leading-edge slats; installation of Rolls-Royce engines; installation of an air-distribution duct to supply fan air to the augmentor flaps and for aileron and fuselage BLC; installation of lateral and directional stability augmentation systems, increased-capacity hydraulic systems and extensive flight test instrumentation; and fixing the landing gear in an extended position. The augmentor flaps, with a constant chord of 3 ft 6 in (1.07 m), are made in four equal span-wise sections, two on each wing, and have a maximum deflection of 75°. They are designed to be efficient ejectors of the fan air and consist of upper and lower segments, each of which is slotted. When extended these flaps deflect the primary jet flow downward and mix it with induced flow coming over the upper wing surface. At the same time, air from above the upper flap is pulled through the slotted surface of the upper flap, and air from below the lower surface of the lower flap is pulled up through that surface's slot, increasing the airflow between the two flap segments. The net effect of this is to combine four different airflows into one jet stream between the two flap segments, increasing both lift and thrust and providing suction-type BLC to prevent or delay airflow separation from the upper flap surface. In early 1974 the proof-of-concept flight phase was completed and the engines were removed for inspection. The second phase of the programme began in September 1974, with a planned 300 hour flight evaluation programme of the STOLAND aerial navigation and terminal guidance system. By the beginning of 1975 a total of 180 flight hours had been accumulated by the aircraft.

(USA)

NASA SUPERCRITICAL WING

First flight (T-2C): November 1970
Purpose: To achieve highly efficient cruise flight near the speed of sound with the supercritical wing
Photo: F-111 with the supercritical wing
Special design features and History: NASA developed a new wing, called the supercritical wing, as a result of wind tunnel studies conducted by Dr Richard T Whitcomb. Simply stated, such a wing utilises an aerofoil shape with a flat top and downward cambered rear section, as compared to the curved top and sloped rear section of a conventional wing. Wind tunnel measured performance showed that the new wing would allow highly efficient cruise flight near the speed of sound. In addition to permitting a substantial increase in cruise speed without increase in power, it could significantly reduce the operational cost of subsonic jet transport aircraft.

When the speed of an aircraft approaches the speed of sound, regions of high supersonic airflow develop, particularly above the wing. These cause severe local disturbances such as shockwaves and boundary layer flow separation, leading to increased drag, severe buffeting and adverse changes in stability. Wind tunnel and analytical studies indicated that the supercritical wing has a potential of permitting subsonic speeds in excess of Mach 0.95 (instead of Mach 0.8 at 35,000 ft) before the adverse effects become significant. The first aircraft to fly with a supercritical wing was a T-2C Buckeye, followed by an LTV Aerospace F-8A Crusader. As fitted to the F-8A, the wing had a span of 43 ft (13.11m), sweepback of 42.24° at quarter-chord, no anhedral or dihedral, incidence of 1°30' at fuselage centreline, aspect ratio of 6.8 and thickness/chord ratio of 11% at root and 7% at tip. Wing area was 274 sq ft (24.45 m^2). As the next stage of development, a supercritical wing was applied to a variable-geometry General Dynamics F-111 and extensively tested under a programme known as Transonic Aircraft Technology (TACT). In this case, the wing was not expected to increase maximum speed, but it was hoped that the aircraft would cruise and manoeuvre at higher speeds without any increase in fuel consumption. Flight testing of the F-111 stopped at the end of 1974, but, after a period during which modifications were made to the aircraft, flights resumed in June 1975. Supercritical wings have also been fitted to the Boeing YC-14 and McDonnell Douglas YC-15 advanced medium STOL transport aircraft for the USAF and other types.

(France) NORD 500

First flight (tethered): July 1968
Purpose: VTOL research aircraft, intended primarily to evaluate the principles of the tilting-duct concept
Power plant: Two Allison 250-C18 turboshaft engines (each 317 shp)
Wing span over ducts: 20 ft 1½ in (6.14 m)
Length: 21 ft 7 in (6.58 m)
Height: 10 ft 2 in (3.10 m)
Diameter of ducts, internal: 5 ft 2 in (1.58 m)
Max T-O weight: 2,760 lb (1,250 kg)
Max level speed, at max T-O weight with developed ducts (estimated): 188 knots (217 mph; 350 km/h)
Accommodation: Pilot only
Special design features and History: The Nord 500 was part of a more general programme which was to include testing of a full-scale "top-fan" system of high static efficiency, with variable expansion of the flow from the duct. The two engines of the N 500 were located side-by-side in the rear part of the fuselage and drove two five-blade ducted pusher propellers through interconnected shafts, running through the stub wings. For vertical take-off and landing, the ducts tilted to a horizontal position, complete with the section of wing to which they were attached. Control in yaw and roll was by differential thrust and tilt, with a "mixing" process during transition. Control in pitch was by collective tilting of the ducts.

The first prototype was completed in the Spring of 1967 and was used as a mechanical and aerodynamic testbed for ground testing only. The second prototype, to which the data above apply, made its initial tethered flight in July 1968. The first year of testing was carried out with the aircraft in tethered hovering flight; and was followed by dynamic stability tests in wind tunnel and simulator. In its initial form, the Nord 500 was intended only for hovering and transition flight trials. Stage 2 development planned to utilise controlled and variable expansion of the flow from the ducts. A new version of the N 500 was designed, with hydraulically-operated controls and powered by 370 hp Allison 250-C20 turboshafts; but development was subsequently suspended.

(France)

NORD GERFAUT

First flight: 15 January 1954
Purpose: Delta-wing research aircraft to collect data for high-speed fighter design
Photo: Gerfaut IA
Data: Gerfaut II
Power plant: One SNECMA Atar 101G turbojet engine with Nord afterburner (9,700 lb; 4,400 kg st)
Wing span: 21 ft 10½ in (6.68 m)
Length: 32 ft 9¾ in (10.00 m)
Height: about 13 ft 5½ in (4.10 m)
Weight loaded: 11,575 lb (5,250 kg)
Max level speed at 39,375 ft (12,000 m): Mach 1.3
Service ceiling: 52,500 ft (16,000 m)
Endurance: 50 mins
Accommodation: Pilot only
Special design features: Thin delta wings with 58°25′ sweepback on leading-edge. Aspect ratio 2.16. Thickness/chord ratio 5½%. Swept vertical tail with high-mounted variable-incidence tailplane of delta planform. Large blister between fin and fuselage, housing brake parachute. Inward-retracting main wheels. Air intake in nose of fuselage, with a "through" duct straight to engine. Pilot's cockpit and fuel tanks situated above this air duct and built as a separate superstructure. Knife-edge windscreen.
History: The Nord Gerfaut IA was the first high-powered jet delta-wing aircraft to fly in France. On 3 August 1954 it became the first aircraft in Europe to exceed Mach 1 in level flight without the use of an afterburner or rocket power, although an afterburner was subsequently fitted. This aircraft was followed by the Gerfaut IB, with larger wings and other improvements, which exceeded the speed of sound in level flight for the first time on 11 February 1955. The further improved Gerfaut II first flew on 17 April 1956. Although it was similar to the Gerfaut I in some constructional features, 80% of its components were redesigned. It was flown at supersonic speed on many test flights, and was used to test interception radar equipment at high speeds and high altitudes. On 16 February 1957 the Gerfaut II established a number of time-to-height records from a standing start, including climb to a height of 6,000 m in 1 min 17 sec, 9,000 m in 1 min 34 sec, 12,000 m in 2 min 18 sec and 15,000 m in 3 min 56 sec. On 28 February it set up a further record by climbing to 3,000 m in 51·2 sec.

(France)

NORD 1500 GRIFFON

First flight: 20 September 1955
Purpose: Experimental aircraft to flight test a new airframe design embodying a combination turbojet-ramjet propulsion unit
Power plant: Fitted originally with one SNECMA Atar 101F turbojet engine with afterburner. Griffon II fitted with one SNECMA Atar 101 E3 turbojet engine (7,716 lb; 3,500 kg st), mounted inside a 54 in (1.37 m) diameter integral ramjet of Nord design, just forward of the ramjet burners
Wing span: 26 ft 7 in (8.10 m)
Length: 47 ft 8½ in (14.54 m)
Height: approx 16 ft 5 in (5.00 m)
Wing area: 344.5 sq ft (32 m²)
Weight loaded: 14,839 lb (6,745 kg)
Max level speed: over Mach 2.1
Accommodation: Pilot only
Special design features and History: The prototype Nord 1500-01 Griffon I made its maiden flight powered by an orthodox turbojet engine (Atar 101F). After completion of the first phase of testing, the airframe was modified to take the advanced propulsion unit for which it was designed, as described above. The ramjet was equipped with fuel pumps able to deliver 8,800 Imp gallons/hr (40,000 litres/hr) and a fuel regulation system consisting of a flow output regulator controlled by an electronic computer, with a limiting device for mixture control according to altitude, speed and throttle position. Although basically the same aircraft, the re-engined model was known as the 1500-02 Griffon II. It flew for the first time on 23 January 1957, and on 17 May exceeded Mach 1 in level flight with its ramjet ignited, though not using the full power of its combination power-unit. From that time the Griffon made over 200 flights at progressively-higher speeds. On 13 October 1959, it achieved Mach 2.19 (1,257 knots; 1,448 mph; 2,316 km/h) at 50,000 ft (15,250 m), and tests continued under a USAF research contract.

The Griffon had 60° delta wings of 4.5% thickness/chord ratio, fitted with elevons. Low pressure tyres, flaps and a braking parachute enabled it to operate from all types of airfield. A fixed forward stabiliser, which gave the Griffon a tail-first layout, ensured accurate longitudinal trim at all speeds, including the transonic zone. There were also airbrakes on the fuselage nose-section.

At Mach 2, the ramjet gave over 80% of the total thrust, which was 9,200 lb (4,170 kg) at 50,000 ft (15,250 m) and 5,500 lb (2,500 kg) at 60,000 ft (18,300 m). Higher speeds were prevented by thermal limitations on the airframe.

(France)

NORD 1601

First flight: 24 January 1950
Purpose: To investigate the stability of swept wings, the effects of sweepback on high-lift devices, and other aerodynamic problems at high subsonic speeds
Photo: Nord 1601
Power plant: Two Rolls-Royce Derwent 5 turbojet engines (3,500 lb; 1,587 kg st)
Wing span: 40 ft 10¾ in (12.463 m)
Length: 38 ft 9½ in (11.823 m)
Height: 15 ft 5 in (4.700 m)
Wing area: 325.3 sq ft (30.22 m²)
Weight loaded: 14,770 lb (6,700 kg)
Max level speed: 539 knots (621 mph; 1,000 km/h)
Service ceiling: over 39,375 ft (12,000 m)
Accommodation: Pilot only
Special design features: Thin laminar-flow wings, swept at 33° at quarter-chord. 8° dihedral from wing roots. Leading-edge slots, each in two independent sections. Slotted trailing-edge flaps. Short-span ailerons outboard of flaps operated in conjunction with spoilers. Spoilers could also be raised simultaneously and independently of ailerons to serve as dive brakes. Engines in nacelles on each side of fuselage. Ejection seat. Provision for technical observer's seat aft of pilot.
History: Originally to have been powered by two Rateau SRA-1 turbojet engines, the Nord 1601 first flew at the beginning of 1950 with Derwents. Development of the Rateau engine had begun in France during the German occupation in the Second World War. After the war, with the assistance of the French Air Ministry, construction of the first Rateau turbojet was started. This engine, the SRA-1, completed its acceptance tests successfully in February 1947, but by then had been surpassed by later British and American turbojets. This resulted in the switch to Derwents to power the Nord 1601. An all-weather fighter version of the design, designated Nord 1600, was projected but never built.

(USA)

NORTH AMERICAN F-107A

First flight: 10 September 1956
Purpose: Experimental advanced fighter-bomber development of the F-100 Super Sabre
Power plant: One Pratt and Whitney J75 turbojet engine (24,500 lb; 11,115 kg st with afterburning)
Wing span: 36 ft 7 in (11.15 m)
Length: 60 ft 10 in (18.54 m)
Height: 19 ft 8 in (5.99 m)
Max speed: about Mach 2.2
Accommodation: Pilot only
Special design features: Design evolved from that of F-100 Super Sabre. Bifurcated air intakes on top of fuselage behind cockpit, with central knife-edge which created inclined shock-waves to slow airflow entering engine at supersonic speeds. One-piece all-moving fin; spoiler-type ailerons.
History: The F-107A, known originally as the F-100B, was to have been an advanced fighter-bomber development of the F-100 Super Sabre. Three F-107As were built, the first flying in September 1956, and exceeded Mach 2.0 on many occasions during subsequent test flights at Edwards AFB, California. Further development was cancelled as an economy move in 1957, but two of the aircraft were turned over to the NACA for high-speed flight research. The third was placed in the Air Force Museum at Wright-Patterson AFB, Dayton, Ohio.

(USA)

NORTH AMERICAN X-15A

First flight (powered; second aircraft): 17 September 1959
Purpose: High-performance research aircraft, used to furnish data with regard to heating, stability, control and the problems of re-entry into the atmosphere
Data: X-15A-2
Power plant: One Thiokol (Reaction Motors) XLR99-RM-2 single-chamber throtteable liquid-propellant rocket motor (57,000 lb; 25,855 kg st at 45,000 ft; 13,700 m)
Wing span: 22 ft 0 in (6.70 m)
Length overall: 52 ft 5 in (15.98 m)
Max launching weight: 50,914 lb (23,095 kg)
Max attained speed: Mach 6.72 (3,937 knots; 4,534 mph; 7,297 km/h) km/h)
Height attained on 22 August 1963: 354,200 ft (107,960 m), or 67.08 miles
Special design features: Aircraft made largely of titanium and stainless steel. Airframe covered with an "armour skin" of Inconel X nickel alloy steel, to withstand temperatures ranging from +1,200°F to −300°F. (Temperatures far beyond these were reached after the entire airframe of the X-15A-2 had been coated with Emerson Electric T-500 ablative material). Rounded "hot-nose" designed to sense the angles of attack and side-slip during exit and re-entry phases of hypersonic flight in upper atmosphere. Wings of thin (5%) section, with 2.5 aspect ratio. Pitch and roll control in atmosphere provided by all-moving tailplane, the two parts of which could be moved together or differentially. Dorsal and ventral fins of wedge section, containing small split airbrakes. Upper fin pivoted for directional control; lower fin jettisonable before landing. Control at high altitude by means of 12 jet nozzles of the HTP rocket type, four in wingtips and eight in nose.
History: The X-15A was flown faster and higher than any other manned aeroplane. In 1955 North American was awarded a USAF/USN/NACA contract for three manned research aircraft with a design speed of at least Mach 7 and capable of reaching an altitude of at least 264,000 ft (80,500 m). Initial flight tests with each of the first two X-15As were made with two Reaction Motors LR11-RM-5 rocket motors (each 8,000 lb; 3,630 kg st), as the XLR99 engine was not then ready for service. The first X-15A flew for the first time under the wing of its B-52 mother-plane on 10 March 1959, but was not released. On 8 June 1959 the first unpowered free flight was made. From 1961, numerous flights set up successively higher unofficial international performance records. No 2 was involved in an accident on 9 November 1962 and was rebuilt in a new form, with additional propellants in large external tanks, for even more advanced research. In this form it first flew on 28 June 1964, and subsequently reached a yet-unbeaten speed of Mach 6.72. On 15 November 1967, X-15A-3 was lost in a crash near Johannesburg, California, and the X-15 programme was terminated in November 1968 after a total of 199 flights had been made.

(USA)

NORTH AMERICAN XB-70A VALKYRIE

First flight: 21 September 1964
Purpose: Mach 3 aerodynamic test aircraft (planned originally as strategic bomber project)
Power plant: Six General Electric YJ93-GE-3 turbojet engines (each 31,000 lb; 14,060 kg st with afterburning)
Wing span, tips spread: 105 ft 0 in (32.00 m)
Foreplane span: 28 ft 9¾ in (8.78 m)
Length overall: 196 ft 0 in (59.74 m)
Designed cruising speed: about Mach 3
Range: about 6,600 nm (7,600 miles; 12,230 km) unrefuelled
Accommodation: Crew of two for research purposes
Special design features: Delta wings of very thin section, with slight camber at root. Aspect ratio 1.751. Chord 117 ft 9 in (35.89 m) at root. Anhedral over entire span and slight washout twist. Sweepback on leading-edge 65°34′. Entire wings covered with brazed stainless steel honeycomb sandwich panels, welded together (similar construction for ducts, part-fuselage and tail unit). Wingtips folded down hydraulically to an angle of 25° for low-altitude supersonic flight and to 65° for high-altitude Mach 3 cruising flight, to improve stability and manoeuvrability. Total of 12 elevons. Three-axis stability augmentation system fitted. Large canard foreplane of very thin section, fitted with trailing-edge flaps. Rectangular-section power plant duct under wings, incorporating Hamilton Standard inlet control system. Two vertical tail-fins and large rudders.
History: The Valkyrie was designed as a strategic bomber to replace the B-52 Stratofortress in service with the USAF in the mid-1960s. Unlike earlier supersonic bombers, the operational B-70 was designed to travel the entire distance to the target and back at Mach 3, carrying nuclear and conventional weapons. After a succession of policy changes, it was decided in 1963 to build only two XB-70As for research. The first flight was made from Palmdale, California, to Edwards AFB. Mach 1 was exceeded for the first time during the third flight, on 12 October 1964, and Mach 3 was attained for the first time on 14 October 1965. The second XB-70A flew at Mach 3 for 32 minutes on 19 May 1966, but was lost on 8 June when an F-104 chase aircraft collided with it at the end of a 2¼-hour flight. At the time of the accident, the two aircraft had flown 95 times, logging 184 hours 10 minutes in the air. The first XB-70A went on to complete 71 flights by 20 February 1968. Meanwhile, management of the programme had been taken over by NASA on 25 March 1967. It was terminated in 1969.

NORTHROP B-35 and YB-49

(USA)

First flight (XB-35): 25 June 1946
Purpose: Flying-wing heavy bombers
Photo: YB-49
Data: B-35
Power plant: Four Pratt and Whitney Wasp-Major (two R-4360-17 and two R-4360-21) radial piston engines (each 3,000 hp), with two single-stage General Electric turbosuperchargers on each engine
Wing span: 172 ft 0 in (52.43 m)
Length: 53 ft 1 in (16.18 m)
Height: 20 ft 1 in (6.12 m)
Wing area: 4,000 sq ft (371.6 m^2)
Weight loaded: up to 209,000 lb (94,800 kg)
Useful load: 73,000 lb (33,112 kg)
Max level speed (YB-49): over 434 knots (500 mph; 805 km/h)
Accommodation: Crew of seven
Special design features: Cantilever wing of aluminium alloy, constructed in one piece, straight-tapered and sweptback. Max chord (on centre-line) 37 ft 6 in (11.43 m). Tip chord 9 ft 4 in (2.84 m). Dihedral on chord line 1°. Leading-edge sweepback 27°57′. Drag-inducing double-split flaps at wingtips for directional control; elevons between them and outer engines. Fixed wingtip slots in leading-edge, which opened only at speeds approaching the stall. Fuselage nacelle built around centreline of wing for the crew. Engines drove Hamilton Standard four-blade reversible-pitch pusher propellers. Two electrically-operated four-gun turrets, one above and one below wing. Four electrically-operated remotely-controlled two-gun turrets, one above and one below each wing between outer engines and wingtips. Fire control blister towards end of central nacelle.
History: Preliminary data on a long-range bomber of this configuration were submitted to the USAAF in September 1941; with the assistance and co-operation of the Wright Field Engineering Division, actual design work began in 1942. Four twin-engined flying scale models of the XB-35, of roughly one-third size, were built first, to provide research data, under the designation N-9M. Work on the prototype XB-35 began at the Northrop plant in Hawthorne, California, in 1943. This aircraft was completed in 1946 and made its first flight from Hawthorne to Muroc (now Edwards AFB), where it went through its trials. Fourteen development aircraft, designated YB-35s, were ordered by the USAF; two of these were later earmarked for conversion to eight-jet-engined YB-49s and one to a six-jet YRB-49A. The first YB-49 flew on 21 October 1947. In 1948 an order was placed for 30 B-49s, but this was cancelled in January 1949. The YRB-49A, which flew for the first time on 4 May 1950, was tested at Edwards AFB and became the sole survivor of the series.

(USA)

NORTHROP X-4 BANTAM

First flight: 15 December 1948
Purpose: To investigate the stability and flight characteristics of aircraft of sweptback tail-less configuration at subsonic speeds
Power plant: Two Westinghouse J30-WE-7-9 turbojet engines (each 1,600 lb; 726 kg st)
Wing span: 26 ft 10 in (8.18 m)
Length: 23 ft 3 in (7.10 m)
Height: 10 ft 8 in (3.25 m)
Wing area: 200 sq ft (18.58 m^2)
Designed max weight: 6,888 lb (3,124 kg)
Max speed: about Mach 0.89 at 30,000 ft (9,150 m)
Range: 365 nm (420 miles; 676 km)
Accommodation: Pilot only
Special design features: Thin-section wings, swept at 41° 33' on the leading edge, and fitted with elevons which served as both ailerons and elevators. No horizontal tail surfaces. High aspect ratio sweptback fin and rudder. Engines located in the wing roots, alongside the fuselage. Integrally-machined panels used extensively in the structure.
History: Two X-4s were built, the first being flown in late 1948 by Charles Tucker. The second aircraft was flown by several pilots, including Charles "Chuck" Yeager of X-1 fame; and in all some thirty flights were made before the aircraft was delivered to the NACA. Under NACA direction, the X-4s were engaged in extensive research flying and a total of about sixty flights were made. The complete research programme was concluded in April 1954. Notwithstanding the small proportions of the pilot's cockpit, an ejection seat was fitted and comprehensive instrumentation was installed. Both X-4s still exist; one at the Air Force Academy at Colorado Springs and the other at the Air Force Museum in Dayton.

(USA)

NORTHROP XP-79B

First flight: 12 September 1945
Purpose: Experimental "flying-wing" tail-less fighter, intended primarily for ramming enemy aircraft, although four 0.5 in guns were carried
Power plant: Two Westinghouse J30 turbojet engines (each 1,150 lb; 522 kg st)
Wing span: 38 ft 0 in (11.58 m)
Length: 14 ft 0 in (4.27 m)
Weight loaded: 8,670 lb (3,933 kg)
Max speed (estimated): 443 knots (510 mph; 821 km/h)
Accommodation: Pilot only
Special design features: "Flying-wing" configuration, with the pilot lying in a prone position in a cockpit on the centreline. Engine on each side of the cockpit, exhausting aft of the wing trailing-edge. Twin fins mounted on top of the engine fairings. Bellows-operated split horizontal control surfaces located in the outer wings, air for their actuation being diverted by control valves in the oval tunnel ducts which formed the wingtips. Four-wheeled landing gear.
History: One of the shortest-lived aircraft, and undoubtedly one of the most unusual, the XP-79B was designed for ramming the tails of enemy aircraft and was constructed of welded magnesium. It had been developed from the MX-324 tail-less glider, which was later fitted with an Aerojet XCAL-200 rocket motor and, in this configuration, flew at 350 mph (563 km/h). Unfortunately, on its first flight, which had a duration of only fifteen minutes, the XP-79B went out of control and was destroyed.

(USA)

NORTHROP/NASA HL-10

First flight (unpowered): 22 December 1966
Purpose: Lifting-body research vehicle
Photo: HL-10 under the wing of its B-52 "mother-plane"
Power plant: One Thiokol XLR-11 rocket engine (8,000 lb, 3,630 kg st), replaced subsequently by three Bell hydrogen-peroxide rockets (each 500 lb; 227 kg st)
Max width: 15 ft 1 in (4.60 m)
Length overall: 22 ft 2 in (6.76 m)
Height: 11 ft 5 in (3.48 m)
Planform area: 162 sq ft (15.05 m^2)
Max launching weight: 9,400 lb (4,265 kg)
Accommodation: Pilot only
Special design features and History: The HL-10 was generally similar in construction to the M2, but had an "opposite" configuration in that the straight edge of the "D" cross-section formed the undersurface of the fuselage. The performance of the HL-10 was also similar to that of the M2, except that its subsonic lift/drag ratio proved to be nearly 4.0. The flying control surfaces of the HL-10 consisted of a rudder on the central fin, split vertically for use as a speed brake, and two very thick blunt-edged elevons. The outer fin trailing-edges each had two surfaces which could be flared for transonic stability. Each elevon also had a movable flap on the upper surface which could be raised for this purpose.

The HL-10 was delivered to NASA on 19 January 1966. After tests in the Ames wind tunnel, and installation of the rocket engine at Edwards AFB, the aircraft made its first unpowered glide flight, being dropped from under the B-52 "mother-plane" at an altitude of approximately 45,000 ft (13,700 m) and speed of Mach 0.8. By the end of 1971 some 37 flight tests had been carried out, of which 25 were powered. The HL-10 attained a maximum altitude of 90,303 ft (27,500 m) and a speed of Mach 1.861 during this test programme. NASA's Flight Research Center subsequently conducted a series of tests with the HL-10 to gather data on the need for space shuttle vehicles to have auxiliary power for use during the landing approach. For this programme, the Thiokol engine was replaced by three Bell rockets which could be fired independently to provide the pilot with three thrust levels.

(USA)

NORTHROP/NASA M2-F2

First flight (unpowered): 12 July 1966
Purpose: Lifting-body re-entry research vehicle
Power plant: One Thiokol XLR 11 rocket engine (8,000 lb; 3,630 kg st)
Max width: 9 ft 7 in (2.92 m)
Length overall: 22 ft 2 in (6.76 m)
Height: 8 ft 10 in (2.69 m)
Planform area: 160 sq ft (14.86 m^2)
Max launching weight: 9,000 lb (4,080 kg)
Accommodation: Pilot only

Special design features and History: Under contract to NASA's Flight Research Center, Northrop built two wingless lifting-body re-entry research vehicles. One, designated M2-F2, was an Ames Research Center concept and represented a more refined metal development of the successful wooden M2-F1 glider. The other was the HL-10, described separately. The M2-F2 had a basic delta planform and was D-shaped in cross section, the straight side of the "D" forming the top. The aircraft was controlled in pitch by a flap on the aft lower surface, and by two flaps on the upper aft surface which were used for pitch trim. These surfaces were also used for roll control. Flaps on the aft outer face of each fin were used for yaw control. Unlike the M2-F1, the F2 was so designed that, after initial unpowered flight trials, it could be fitted with a rocket-engine.

The M2-F2 was delivered to NASA on 15 June 1965. Following tests in the full-scale wind tunnel at Ames Research Center, it made its first unpowered flight after being released from the B-52 "mother-plane" at 45,000 ft (13,700 m). Its pilot made a practice flare-out at about 25,000 ft (7,600 m) and then increased speed to 304 knots (350 mph; 560 km/h) in order to be able to flare out and slow the rate of descent from 250 ft/sec to under 5 ft/sec for touchdown. The four-minute flight ended with a landing on Rogers Dry Lake bed, near Edwards AFB. By the end of 1966, the M2-F2 had completed its initial programme of unpowered tests, having made 14 flights. It was then fitted with the rocket engine. On 10 May 1967, it was damaged in a wheels-up landing at the end of its 16th flight. Purpose of the unpowered flight had been to evaluate the effects of reduction in automatic damping for roll and yaw before the start of powered flight trials. The M2-F2 was completely dismantled, inspected and some components rebuilt. It was then reassembled as the M2-F3, with an additional central fin. In this form, it made its first powered flight on 25 November 1970, achieving Mach 0.8 at 53,000 ft (16,150 m). The final flight was made on 20 December 1972, completing a programme of 25 flights, of which 20 were powered.

(France)

PAYEN PA 49

First flight: 22 January 1954
Purpose: Experimental jet-propelled tail-less flying wing
Power plant: One Turboméca Palas turbojet engine (330 lb; 150 kg st)
Wing span: 16 ft 11 in (5.16 m)
Length: 16 ft 8¾ in (5.10 m)
Height: 8 ft 2½ in (2.50 m)
Weight loaded: 1,433 lb (650 kg)
Max level speed: 270 knots (311 mph; 500 km/h)
Service ceiling: 27,890 ft (8,500 m)
Range: 243 nm (280 miles; 450 km)
Accommodation: Pilot only
Special design features: All-wood construction. Wings with sharp leading-edge sweepback of 50°. Aspect ratio 2.7. Wing-root air intakes. Large triangular fin, the front end of which formed a cockpit fairing, extended over two-thirds of the fuselage length. Non-retractable tricycle landing gear.
History: M. Payen began construction of the Pa 49 in 1951. After satisfactorily completing its preliminary trials at Melun-Villaroche, the aircraft was sent to the Centre d'Essais en Vol, Brétigny, where it completed its first series of tests in August 1954. It was then fitted with a new Fléchair airbrake system. This took the form of a split rudder which was effective as a brake throughout the rudder range. Several developments of the aircraft were projected, including the single-seat Pa 492 and two-seat Pa 495 trainers.

(USA)

PIASECKI 16H-1A PATHFINDER II

First flight (16H-1): 21 February 1962
Purpose: Compound helicopter for high-speed research
Power plant: One General Electric T58 turboshaft engine (1,250 shp)
Diameter of rotor: 44 ft 0 in (13.41 m)
Diameter of propeller: 5 ft 6 in (1.68 m)
Length of fuselage: 37 ft 3 in (11.35 m)
Height to top of rotor head: 11 ft 4 in (3.45 m)
Max T-O weight (STOL): 10,800 lb (4,900 kg)
Max level speed at 8,150 lb (3,697 kg) VTOL T-O weight: 200 knots (230 mph; 370 km/h)
Service ceiling, above weight: 18,700 ft (5,700 m)
Range with max payload, 10% fuel reserve, above weight: 391 nm (450 miles; 725 km)
Accommodation: Crew of two and six passengers
Special design features: Fully-articulated all-metal three-blade main rotor, attached to hub through tension-torsion straps. Anti-torque control by three-blade ducted propeller at tail. Rotor brake. Cantilever low wings of aluminium alloy and honeycomb construction. Aspect ratio 5.3. Combined ailerons and flaps ("Flaperons") fitted. Retractable main landing gear
History: The Pathfinder utilised a ducted propeller (known as a ring-tail), at the rear, to provide directional and anti-torque control by means of vertical vanes in the duct. At a safe height after take-off, increased power was put into the ducted propeller for forward propulsion. In cruising flight the small-span fixed wings off-loaded the rotor.

In its original form, as the five-seat PiAC 16H-1, the Pathfinder was developed as a private venture. Powered by a 550 shp P & W PT6B-2 shaft-turbine engine, it logged a total of 185 hours flying time, during which speeds of up to 148 knots (170 mph; 273 km/h) were attained. In 1964, under a US Army and Navy contract, PiAC began modifying the Pathfinder to make it capable of attaining speeds of up to 200 knots (230 mph; 370 km/h). Now designated 16H-1A Pathfinder II, it had a new engine, new drive system and propeller, a larger rotor and a lengthened fuselage. Flight testing was resumed on 15 November 1965, when the aircraft made its initial hovering trials. By May 1966, it had logged more than 40 flying hours under a joint Army/Navy programme, including flight at forward speeds of up to 195 knots (225 mph; 362 km/h). At that time, approximately 20 autorotative flights had been made, at speeds between 39 and 100 knots (45-115 mph; 75-185 km/h). The 16H-1A was later re-engined with a 1,500 shp General Electric T58-GE-5 turboshaft engine and was redesignated Model 16H-1C.

(USA)

PIASECKI VZ-8P (B) AIRGEEP II

First flight: Summer of 1962
Purpose: Experimental vertical take-off aircraft and land vehicle
Power plant: Two Turboméca Artouste IIC (modified) shaft-turbine engines (each 530 shp)
Length: 24 ft 6 in (7.47 m)
Width: 9 ft 3 in (2.82 m)
Height: 5 ft 10 in (1.78 m)
Max T-O weight: 4,800 lb (2,177 kg)
Accommodation: Seating for five
Special design features and History: The Airgeep II (Model Pa-59H) was an improved version of the Model 59K Airgeep I. It was an experimental vertical take-off aircraft which, like its predecessor, was developed under a contract from the US Army Transportation Research Command. Its two horizontal opposite-rotating three-blade ducted rotors were driven by the Artouste engines, and it had powered wheels for self-contained ground mobility and for extended range when conditions permitted driving overland. The wheels were arranged like a non-retractable tricycle landing gear.

The pilot's flying controls were similar to those of a conventional helicopter, and provided directional control through a series of pivoted vanes mounted under the rotor duct.

The prototype began its programme of ground and flight tests in mid-1961 in stripped configuration, and made its first free flight in the Summer of 1962.

On military observation, liaison and other duties, this type of aircraft was intended to fly close to the ground, taking advantage of natural cover. However, the VZ-8P (B) was in no way dependent on ground effect, and was capable of flying at a height of several thousand feet.

(USA)

REPUBLIC XF-91

First flight: 9 May 1949
Purpose: Experimental high-speed interceptor fighter
Power plant: One General Electric J47-GE-3 turbojet engine with afterburner (5,200 lb; 2,360 kg st) and a Reaction Motors XLR11-RM-9 auxiliary rocket motor (6,000 lb; 2,720 kg st)
Wing span: 31 ft 3 in (9.53 m)
Length: 46 ft 8 in (14.22 m)
Height: 18 ft 1 in (5.51 m)
Weight loaded: 30,000 lb (13,608 kg)
Max level speed: over Mach 1
Accommodation: Pilot only
Special design features: Variable-incidence sweptback wings, with inverse taper and thickness, giving greater chord and depth at the tips than at the roots. Tandem main-wheel landing gear units which retracted outward into the thicker and wider portions of the wings near the tips. The four tubes of the rocket motor were disposed two above and two below the jet orifice. Sweptback tail surfaces, with the tailplane about one-third of the way up the high tail-fin.
History: Two prototypes of the XF-91 were built, the first flying on jet power alone in May 1949. The variable incidence permitted high angle of attack for take-off and landing; the inverse taper, combined with leading-edge slots, reduced wingtip stalling tendency at low speeds. In December 1952, at Edwards Air Force Base, California, the XF-91 exceeded Mach 1, using full jet and rocket power, so becoming the first American combat-type aircraft to fly faster than sound in level flight.

(Germany)

First flight: October 1970
Purpose: Experimental aerofoil boat, built to provide data for larger craft
Power plant: One Nelson H63-CP four-cylinder piston engine (48 hp, derated to 40 hp)
Wing span: 19 ft 3¾ in (5.89 m)
Length: 27 ft 8 in (8.43 m)
Height: 6 ft 9½ in (2.07 m)
Weight loaded: 760 lb (345 kg)
Accommodation: Pilot only
Special design features: Anhedral wings of reversed delta planform, with outboard integral floats and dihedral wingtips with control surfaces. Rectangular tailplane mounted on top of high fin and rudder. Engine supported on struts above fuselage. Airframe of special glassfibre sandwich construction, with a core of tubular or foam plastics.
History: The Aerofoil Boat concept was begun in the United States by Dr. A. M. Lippisch, and was first tested in the Collins X-112 craft of the early 1960s. Since 1967 further development of the concept has been undertaken by Rhein-Flugzeugbau GmbH, with assistance from the German Federal government, under the designation X-113, as a preliminary to experiments with a larger craft of the same type. Dr. Lippisch remains the technical and scientific manager of the project. The present X-113 Am underwent its first airworthiness test from Lake Constance in October 1970, and flight characteristics and performance have proved satisfactory. Since that time, tests carried out in the North Sea have shown that the Aerofoil Boat can be operated successfully in rough water conditions.

RFB (LIPPISCH) X-113 Am AEROFOIL BOAT

(USA)

RYAN VZ-3RY VERTIPLANE

First flight: 29 December 1958
Purpose: V/STOL research aircraft
Power plant: One Lycoming T53-L-1 shaft-turbine engine (1,000 shp)
Wing span: 23 ft 5 in (7.14 m)
Length: 27 ft 8 in (8.43 m)
Height: 10 ft 8 in (3.25 m)
Weight loaded: approx 2,600 lb (1,179 kg)
Accommodation: Pilot only
Special design features and History: The Vertiplane, built for the US Army under the technical direction of the Office of Naval Research, was a simply-constructed V/STOL research monoplane of fairly conventional high-wing configuration but with very extensive double retractable wing flaps which extended far below and to the rear of the wing trailing-edge. The single engine drove two wing-mounted Hartzell three-blade wooden propellers of large diameter. Large endplates at the wingtips provided structural support for the flaps and confined the slipstream to the flap span for maximum efficiency. Conventional stick and rudder pedal controls were fitted in the cockpit to operate the rudder, elevator and variable-incidence tailplane, and the spoilers which were inset in the upper surface of each wing ahead of the flaps, taking the place of the usual ailerons. A universally-jointed jet-deflection nozzle at the rear of the tailpipe from the engine was intended to ensure adequate control during hovering flight.

The prototype began its taxiing trials on 7 February 1958, and, after prolonged ground testing, including three months of tests in the full-scale low-speed wind tunnel of the NASA Ames Laboratory at Moffett Field, it made its first take-off as a conventional aeroplane in late 1958. In a subsequent six-week test programme, it proved able to make near vertical take-offs at a ground speed of 22 knots (25 mph; 40 km/h) after a run of 30 ft (9 m). It hovered at zero air speed at altitudes between 100 and 3,700 ft (30-1,125 m), and accomplished transitions from hovering to forward flight. After making 21 successful flights and being delivered to NASA, the VZ-3RY was virtually destroyed in an accident. However, NASA decided to rebuild the aircraft for further testing.

(USA)

RYAN X-13 VERTIJET

First flight: 10 December 1955
Purpose: VTOL research aircraft
Power plant: One Rolls-Royce Avon turbojet engine (10,000 lb; 4,535 kg st)
Wing span: approx 21 ft 0 in (6.40 m)
Length: approx 24 ft 0 in (7.32 m)
Height: approx 15 ft 0 in (4.57 m)
Weight loaded: approx 7,500 lb (3,401 kg)
Thrust/weight ratio: 1.3: 1
Accommodation: Pilot only
Special design features: High-set delta wing with small fins at tips. Large vertical fin and rudder. No landing gear, aircraft taking off instead from a mobile ground servicing trailer, the bed of which was raised hydraulically to a vertical position so that the aircraft was suspended in a nose-up attitude by a hook under the forward fuselage. Pivoted pilot's seat.
History: The X-13 was a tail-sitting VTOL research aircraft. Ryan began research into the possibility of such an aircraft in 1947, by mounting an Allison J33 engine on a horizontal test stand and investigating methods of varying the jet thrust for directional control. Next, the J33 was mounted in a vertical test rig, and finally was fitted with a cockpit, controls and delta wings for piloted tests. As a result of this successful programme, Ryan received a USAF contract to build the X-13 in 1953, and the first prototype flew while fitted with a temporary fixed tricycle landing gear to permit normal horizontal take-offs and landings. This landing gear was subsequently removed and replaced by a temporary tubular structure to allow the aircraft to be tested in the tail-sitting attitude. The first such flight was made on 28 May 1956, followed by simulated hook-ons to a nylon rope stretched between two upright steel towers. A second prototype made the first transition from horizontal flight to hovering flight on 28 November 1956. In its fully developed form, the X-13 could be hovered and manoeuvred close to the ground. During take-off, landing and hovering flight, it was controlled by deflection of its exhaust and thrust variations. Conventional control surfaces were provided for horizontal flight. The first complete sequence of vertical take-off, transition to horizontal cruising flight and return to vertical landing was accomplished on 11 April 1957. A complete test programme was carried out with the X-13s.

(USA)

RYAN XF2R-1

First flight: November 1946
Purpose: Experimental fighter, used to gain experience with the XT31-GE-2 (TG-100) engine
Power plant: One General Electric XT31-GE-2 turboprop engine (2,200 shp and 600 lb thrust) and one General Electric J31-GE-3 turbojet engine (1,600 lb; 726 kg st)
Wing span: 40 ft 0 in (12.19 m)
Length: 36 ft 0 in (10.97 m)
Height: 14 ft 0 in (4.27 m)
Wing area: 275 sq ft (25.54 m²)
Weight loaded: 11,000 lb (4,990 kg)
Max level speed: about 434 knots (500 mph; 805 km/h)
Accommodation: Pilot only
Special design features: Basically a Ryan FR-1 Fireball fighter with a longer fuselage to house the new forward engine. Dorsal extension to the fin.
History: The TG-100 was the first turboprop engine to be designed, built and flown in the United States. The first example was completed in 1943; another was given its initial flight test on 21 December 1945, in a Consolidated Vultee XP-81. By the Summer of 1948, well over 100 flights had been completed with the XP-81 and the Ryan XF2R-1. The latter aircraft was a development of the Fireball fighter and, while retaining the aft-mounted turbojet engine, was fitted with a forward TG-100 turboprop in place of the Fireball's Wright Cyclone piston engine. As an experimental fighter it was intended that only the turboprop engine should be used for cruising flight, and that the rear-mounted turbojet would come into operation for combat boost. An altitude of 39,100 ft (11,900 m) was reached by the XF2R-1 during its initial flight tests.

(USA) RYAN XV-5A and XV-5B

First flight (XV-5A): 25 May 1964
Purpose: "Fan-in-wing" VTOL research aircraft
Data: XV-5A
Power plant: Two General Electric J85-GE-5 turbojet engines (each 2,658 lb; 1,205 kg st)
Wing span: 29 ft 10 in (9.09 m)
Length overall: 44 ft 6¼ in (13.56 m)
Height: 14 ft 9 in (4.50 m)
Wing area, gross: 260.32 sq ft (24.19 m²)
Max T-O weight (VTOL): 12,300 lb (5,580 kg)
Max level speed (estimated): 475 knots (547 mph; 880 km/h)
Service ceiling (estimated): 40,000 ft (12,200 m)
Range with max fuel, pilot only, at 40,000 ft (12,200 m), (estimated): 868 nm (1,000 miles; 1,610 km)
Accommodation: Crew of two
Special design features: Mid-positioned wings. Aspect ratio 3.419. Chord 12 ft 1 in (3.68 m) at root, 3 ft 7 in (1.09 m) at construction tip. Thickness/chord ratio 10% at root, 11% at tip. Dihedral 0° inboard, 4° on outer wings. Incidence 0°. Sweepback at quarter-chord 15° 32′ inboard, 28° 20′ on outer wings. Single-slotted aluminium flaps. Trim tab in each aileron. Fan between spars in each inner wing. Conventional fuselage structure, with fan in nose. Longitudinally-hinged air inlet doors above fan and reverser doors under fan. Variable-incidence tailplane mounted on top of fin. Thrust spoilers aft of tailpipes enabled engines to be opened up to full power with the aircraft stationary, before diverter valve was opened.
History: In November 1961, after two years of research by General Electric Co and Ryan had proved the practicability of the "fan-in-wing" VTOL principle, the US Army Transportation Research Command ordered two XV-5A prototype aircraft to flight test the concept. General Electric was appointed prime contractor and was responsible for supplying the lift and propulsion systems; Ryan designed and built the aircraft. In the flight test/research programme that followed, these two prototypes between them made a total of 338 flights in all modes of operation and accumulated 138 flying hours between May 1964 and October 1966. One of the prototypes was lost in April 1965; the second was damaged in October 1966. The second aircraft was subsequently rebuilt in modified form for NASA and designated XV-5B. Modifications included a new main landing gear, revised cockpit layout, and removal of the thrust spoiler mechanism. Preliminary flight tests of the XV-5B began on 24 June 1968 and the aircraft was, after additional testing, handed over to NASA for use in its aeronautical research programme.

(Sweden)

SAAB 210 DRAKEN

First flight: December 1951
Purpose: To test the "double-delta" wing configuration at subsonic speeds
Power plant: One Armstrong Siddeley Adder axial-flow turbojet engine (1,050 lb; 475 kg st)
Wing span: about 16 ft 0 in (4.88 m)
Length: over 20 ft 0 in (6.10 m)
Max level speed: over 347 knots (400 mph; 644 km/h)
Accommodation: Pilot only
Special design features: The wing, of extremely low aspect ratio, had a planform made up of two triangles, known as a "double-delta" form. The inner section of the wing was swept at 80° and the outer at 57°. There was no conventional tailplane, elevator control being combined with the ailerons. Retractable tricycle landing gear, the main units of which could, for test purposes, be moved forward or backward. The C.G. could also be varied in flight by pumping liquid between trim tanks in the nose and tail. Air intakes, positioned in wing roots, extended initially to nose of aircraft but were later cut back to a position level with forward end of cockpit enclosure. Drag parachute. Wide range of special instrumentation. Ejection seat.
History: With an approximately half-scale version of the wing used later on the Saab 35 Draken single-seat fighters of the Swedish Air Force, the Saab 210 completed over 100 test flights by mid-1952.

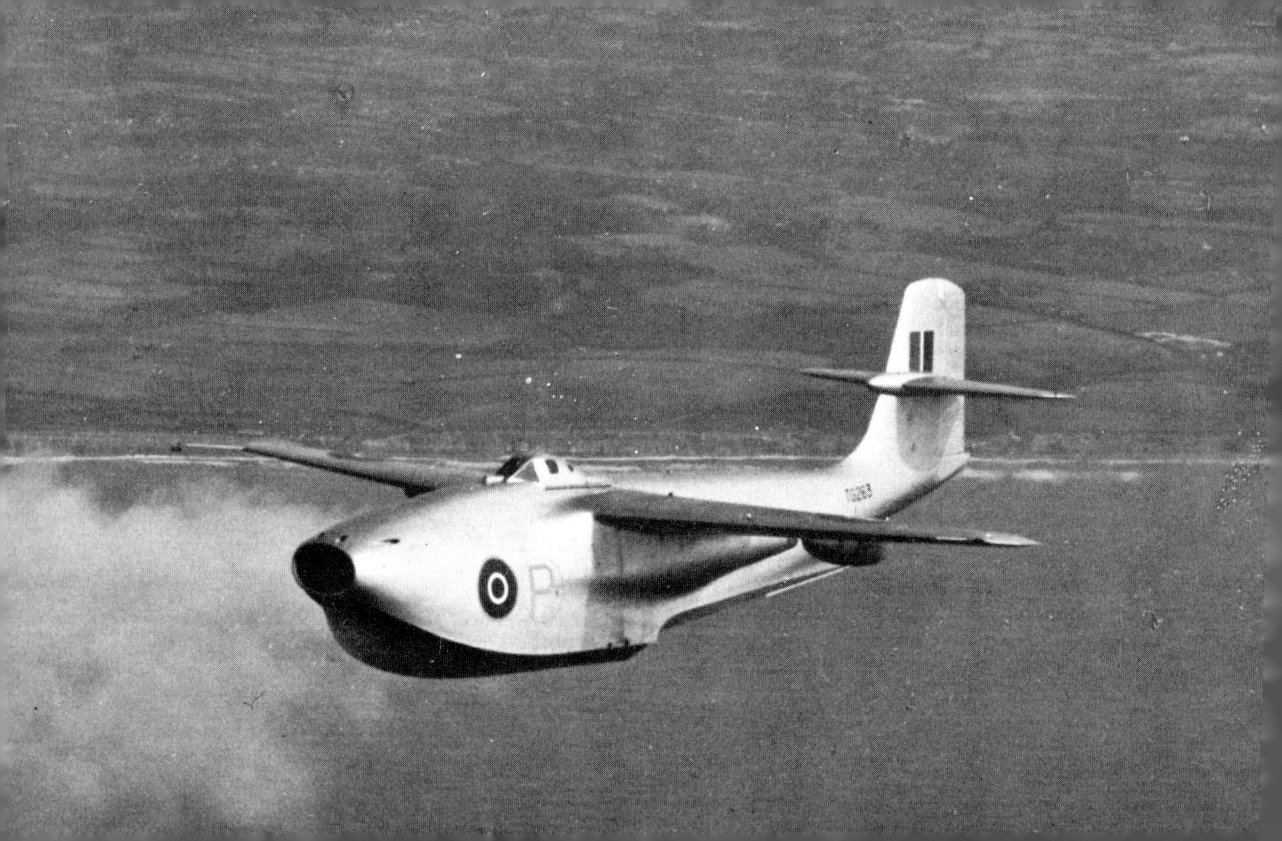

(UK)

SAUNDERS-ROE SR.A/1

First flight: 16 July 1947
Purpose: Experimental jet fighter flying-boat
Power plant: Two Metropolitan-Vickers Beryl M.V.B.I turbojet engines (each 3,250 lb; 1,474 kg st)
Wing span: 46 ft 0 in (14.02 m)
Length: 50 ft 0 in (15.24 m)
Height: 16 ft 9 in (5.11 m)
Wing area: 415 sq ft (38.55 m^2)
Weight loaded: 16,255 lb (7,373 kg)
Max level speed: 445 knots (512 mph; 824 km/h)
Accommodation: Pilot only
Special design features: High-speed aerofoil section used for wings. Semi-retractable stabilising floats between hull and wingtips. Two jet engines mounted side by side in hull. Oval air intake in nose and circular jet outlets on each side of hull aft of wings. Retractable pick-up device under planing bottom for automatic mooring. Martin-Baker ejection seat. Four 20mm cannon in nose.
History: Designed to Specification E.6/44, the SR.A/1 was the world's first jet flying-boat and the first flying-boat to exceed 500 mph (805 km/h) in level flight. It was thought that such a fighter would make the RAF less vulnerable to airfield attack. Three prototypes were built, the second and third being powered by engines of 3,500 lb (1,587 kg) st and 3,850 lb (1,746 kg) st respectively. After a period of extensive trials, research continued with only one of the prototypes. The project was subsequently abandoned, as it became clear that the large hull diminished both speed and manoeuvrability compared with land-based jet fighters.

(UK)

SAUNDERS-ROE S-R.53

First flight: 16 May 1957
Purpose: Experimental "mixed-power" interceptor
Power plant: One de Havilland Spectre variable-thrust long-life rocket motor (about 8,000 lb; 3,630 kg st) and one Armstrong Siddeley Viper turbojet engine (about 1,750 lb; 795 kg st)
Wing span: 25 ft 1¼ in (7.65 m)
Length: 45 ft 10 in (13.97 m)
Height: 10 ft 10 in (3.30 m)
Weight loaded (estimated): 12,780 lb (5,797 kg)
Max level speed (estimated): over Mach 2.4
Accommodation: Pilot only
Armament: Two Firestreak air-to-air guided weapons carried at wingtips
Special design features: Cropped-delta wings with full-span leading-edge flaps. All-moving delta tailplane, with rounded tips, mounted at top of slightly swept fin. Mixed power unit, the turbojet being mounted in the centre section of the fuselage, with the Spectre located in the rear fuselage beneath the Viper tailpipe. Two small air intakes for the Viper located behind cockpit canopy. Both units used kerosene, with high test peroxide serving as the oxidant for the rocket motor.

History: The S-R.53, first British manned aircraft to employ the "mixed-power" formula, was built under Ministry of Supply contract. The first of two prototypes flew in May 1957. From the S-R.53 was evolved the higher-powered S-R.177 which was designed to meet a naval requirement. The construction of a number of prototypes was well advanced when, in December 1957, the contract was cancelled in accordance with the policy outlined subsequently in a British Government White Paper on Defence, which envisaged replacement of manned interceptors by surface-to-air missiles.

(UK)

SHORT S.B.4. SHERPA

First flight: 4 October 1953
Purpose: To investigate the possibilities of the aero-isoclinic wing
Power plant: Two Blackburn Turboméca Palas turbojet engines (each 330 lb; 150 kg st)
Wing span: 38 ft 0 in (11.58 m)
Length: 31 ft 10 in (9.70 m)
Height: 9 ft 1 in (2.77 m)
Wing area: 230 sq ft (21.37 m²)
Weight loaded: up to 3,268 lb (1,482 kg)
Max level speed: 148 knots (170 mph; 274 km/h)
Accommodation: Pilot only
Special design features: The wing (with aspect ratio of 5.6, anhedral of 1° and sweepback of 42°22′) was an aero-isoclinic structure, with its flexural axis located well aft and a torsional stiffness so adjusted that sufficient nose-up twist was provided to cancel out aerodynamic warping due to flexure. To assist this, wingtip elevons were used in place of the usual trailing-edge controls. These elevons, which were hinged at 30% chord, could be rotated either together or differentially, to act as elevators or ailerons respectively. Wing structure was mainly of spruce and plywood, light alloy being used only in strategic positions. An electrically-actuated anti-balance ᵇ was inset at the inboard end of the trailing-edge of each elevon. Landing flaps were inset in the underside of the wings. No tailplane. Swept tail-fin and rudder. Air intake for engines flush on fuselage deck. Jet efflux pipes on sides of fuselage, aft of wings.
History: The S.B.4 was a private-venture research aircraft built to prove the practical possibilities of the aero-isoclinic wing, which had been proposed by David Keith-Lucas, Short's chief designer. The wing, instead of being stiff, was a relatively flexible structure. The all-moving tips, which comprised about one-fifth of the total wing area, served as both ailerons and elevators and made it possible to dispense with the normal tailplane and elevators. It was expected that the rotating tip controls would prove markedly superior to flap-type controls at transonic speeds and would make the aircraft more manoeuvrable at high altitudes. Although the S.B.4 was not capable of high speeds, the lessons learned from its tests were thought to be applicable to the future design of both civil and military aircraft. The Sherpa was a powered version of the Short S.B.1 and proved easy to handle in flight. On completing its test programme with Shorts, the Sherpa was given to the College of Aeronautics at Cranfield where, barring engine trouble that brought a halt to flying in 1958-60, it was flown until 1964. It was then sent as a ground test aircraft to the Bristol College of Advanced Technology, and finished up at the Skyfame Aircraft Museum at Staverton.

(UK) SHORT S.B.5

First flight: 2 December 1952
Purpose: To investigate problems associated with the low-speed handling characteristics of sweptback wings
Photo: S.B.5 with 60° wing position
Power plant: One Rolls-Royce Derwent turbojet engine (3,500 lb; 1,588 kg st)
Wing span: 25 ft 11¾ in — 35 ft 2¼ in (7.92 m — 10.72 m)
Length: 45 ft 9 in (13.94 m)
Height (high tailplane): 16 ft 7 in (5.05 m)
Max level speed: 287 knots (311 mph; 500 km/h)
Accommodation: Pilot only
Special design features: Scale model of English Electric P.1 prototype (Lightning) fighter. Variable (on ground) degree of sweepback on wings. Variable angle of incidence of tailplane. Variable-position non-retractable mainwheels. Anti-spin and braking parachutes fitted.
History: The S.B.5 research aircraft was designed and built at the request of the Ministry of Supply for the investigation of problems associated with the low-speed handling characteristics of swept wings, in order to provide data for the English Electric P.1 prototype fighter. The S.B.5 was designed so that varying degrees of sweepback could be tested. The tailplane could also be positioned either at the extreme top of the fin or beneath the rear fuselage, and its angle of incidence was variable. The varying degrees of wing sweepback were achieved by fitting alternative components and four configurations could be tested. These were 50° wing sweepback with a high tail unit; 60° sweepback with low tail unit; 60° sweepback with high tail unit; and 69° sweepback with high tail unit. The angle of incidence of the tailplane could be varied in flight from 10° above to 10° below the horizontal. The position of the tricycle landing gear could also be changed to enable each configuration to be tested at various C.G. positions. The S.B.5 completed a series of tests with 50° and 60° sweepback while powered by the Derwent engine, but was re-engined with a Bristol Siddeley Orpheus turbojet to test the 69° configuration, in which form it first flew on 18 October 1960. It eventually finished up at the Empire Test Pilots' School at Farnborough, at which establishment it was used to gain experience in flying short-span aircraft at low speeds.

(UK)

SHORT SC.1

First flight: 2 April 1957
Purpose: VTOL research
Power plant: Five Rolls-Royce RB.108 turbojet engines (each 2,130 lb; 966 kg st)
Wing span: 23 ft 6 in (7.16 m)
Length overall: 29 ft 10 in (9.11 m)
Height: 10 ft 8 in (3.25 m)
Max T-O weight (VTOL): 8,050 lb (3,650 kg)
Max level speed at S/L: 214 knots (246 mph; 396 km/h)
Service ceiling: 8,000 ft (2,440 m)
Range with max fuel and max payload: 130 nm (150 miles; 240 km)
Accommodation: Pilot only
Special design features: Delta wings. Aspect ratio 2.61. Chord 17 ft 0 in (5.18 m) at root. Thickness/chord ratio 10%. No dihedral or incidence. Sweepback on leading-edge 54°. Vertical tail surface only. Non-retractable tricycle landing gear, main legs of which were pivoted and could be moved hydraulically through 15° fore-and-aft, enabling wheels to be positioned further aft of C.G. during vertical landing. Four of the engines, used for jet-lift, mounted vertically in cross-wise pairs, each pair swinging on a transverse axis. (After vertical take-off the engines were inclined rearward to give added forward thrust; before vertical landing they could be directed forward to give a braking effect). Fifth engine exhausted horizontally at the tail for thrust in forward flight. All engines had a compressor bleed, supplying high-pressure air to a common duct which fed control nozzles at wingtips, nose and tail. The second SC.1 was fitted with a Mk 3 auto-stabiliser.
History: Developed to Specification ER143, the SC.1 was the first fixed-wing VTOL aeroplane to be built in the UK. Design and construction began in 1954 and the first of two SC.1s made its first flight at Boscombe Down on 2 April 1957, using normal T-O technique. The second aircraft made the first tethered vertical flight on 6 August 1958. A series of hovering trials in a specially-designed test gantry was then followed by free vertical take-offs from an open platform, the first on 25 October 1958. The maiden vertical flight by the No. 1 aircraft was made on 20 July 1960, and the first complete transition from vertical to forward flight and vice versa was accomplished on 6 April 1960. In May 1961, the SC.1 became the first jet-lift aircraft to fly the English Channel. On 2 October 1963, the second aircraft crashed from a low altitude, killing the pilot, after 81 flights with the Mk 3 auto-stabiliser. It was rebuilt and subsequently joined the first at RAE, Bedford, in 1967. There both aircraft were used for exhaustive research into the landing characteristics of VTOL aircraft.

(USA)

SIKORSKY S-72 (RSRA)

First flight: 1976
Purpose: Rotor System Research Aircraft
Photo: Artist's impression of the S-72
Power plant: Two General Electric T58-GE-5 turboshaft engines (each 1,400 shp), and two optional General Electric TF34-GE-2 turbofan engines (each 8,160 lb; 3,700 kg st)
Rotor diameter: 62 ft 0 in (18.90 m)
Wing span: 41 ft 10 in (12.75 m)
Max T-O weight: 26,200 lb (11,883 kg)
Max level speed: 300 knots (345 mph; 555 km/h)
Accommodation: Crew of three
Special design features: S-61 Sea King power plant, transmission and rotor system for initial testing, although other advanced systems will be installed for evaluation. Conventional straight wings can be installed on the lower fuselage, with the angle of incidence variable from $+15°$ to $-9°$. Fuselage side-mounted turbofan engines can be fitted. Tail unit consisting of swept tail-fin, carrying tail rotor, and straight tailplane. Explosive detachment of rotor blades in an emergency. Fly-by-wire control system with auxiliary mechanical control system.
History: In January 1974, NASA awarded Sikorsky a contract to build two prototype S-72s, together with a pair of wings and two turbofan engines in pods. Responsibility for the initial flight test programme, to cover 80 hours of flying, also falls upon Sikorsky. The helicopters will be delivered for further research by NASA and the US Army in 1977. The S-72s will enable research to be carried out on a wide range of advanced rotor systems, the wing and engine layout even permitting the flight testing of rotors that would otherwise be too small to sustain the helicopter in the air.

(USA)

SIKORSKY XH-59A (MODEL S-69)

First flight: 26 July 1973
Purpose: Research helicopter to test the Advancing Blade Concept (ABC) rotor system
Power plant: One Pratt and Whitney (Canada) PT6T-3 Turbo Twin Pac. Provision for later installation of two P & W J60 turbojet engines in pods on each side of fuselage
Diameter of rotors (each): 36 ft 0 in (10.97 m)
Length of fuselage: 40 ft 9 in (12.42 m)
Height over fins: 12 ft 11 in (3.94 m)
Design speed, with turbojet auxiliary propulsion: 300 knots (345 mph; 555 km/h)
Accommodation: Crew of two
Special design features: Two contra-rotating three-blade main rotors, mounted co-axially. No tail rotor. Fuselage of circular cross-section. Cantilever tail unit like that of fixed-wing aircraft, with twin endplate fins and rudders. Retractable tricycle landing gear.
History: On 7 February 1972 Sikorsky announced that the company was designing and building a research aircraft, designated S-69, to flight test the ABC rotor system, under a contract awarded by Eustis Directorate, US Army Air Mobility Research and Development Laboratory. The contract was later modified to include detail design changes and the construction of two prototypes under the Army designation XH-59A.

The ABC rotor system, consisting of two co-axial counter-rotating rigid rotors, takes advantage of the blades' aerodynamic lift on the advancing side of each rotor disc, and full lift capability of the advancing blade is achieved without any penalty being imposed by the retreating blade. This removes the need for a wing to supplement the rotor. Another advantage of the concept is the elimination of a conventional anti-torque tail rotor and its drive system. The first prototype made its maiden flight in July 1973, as a pure helicopter, but was damaged in a flight accident at Sikorsky's Stratford works in the following month. This led to a number of design changes in the second prototype, including a modified control system. The test programme was resumed on 21 July 1975, when the second prototype made a brief initial flight as a pure helicopter. Eventually, it will have two P & W J60 engines for auxiliary forward thrust in a high-speed compound helicopter configuration.

(France)

SNECMA C.450-01 COLÉOPTÈRE

First flight (free vertical): 6 May 1959
Purpose: VTOL research
Power plant: One SNECMA Atar 101E.V turbojet engine (8,157 lb; 3,700 kg st)
Diameter of wing: 10 ft 6 in (3.20 m)
Length: 26 ft 3½ in (8.022 m)
Weight loaded: approx 6,614 lb (3,000 kg)
Accommodation: Pilot only
Special design features: Annular wing of light alloy construction, consisting of two skins and internal structure (chord 9 ft 10 in; 3.0 m). Retractable foreplanes in fuselage nose. Cruciform fins and rudders to provide directional control in all axes. Four oleo-pneumatic landing legs mounted on trailing-edge of wing; small castoring wheels with rubber tyres. Tilting pilot's seat which could be ejected in an emergency.
History: Initial tests with the "Atar Volant" pilotless and piloted test vehicles proved the ability of a vertically-mounted turbojet to raise a VTOL aircraft safely from the ground, to accelerate it in vertical flight to a speed where it could become airborne like a conventional aircraft, and to return it to the ground in a vertical descent. SNECMA then built a prototype research aircraft around this type of power plant. Known as the C.450-01 Coléoptère, this prototype was basically similar to the C.400 P-3 piloted "Atar Volant", but was fitted with an annular wing to permit transition into horizontal flight. The airframe, built by the Nord company in its Chatillon-sous-Bagneux works, was intended for tests at subsonic speeds. Directional control at take-off and landing was by pneumatic deflection of the main jet efflux; directional control during normal horizontal flight was by four swivelling fins equally spaced around the rear of the annular wing. Under an agreement signed in 1958, the Federal German Ministry of Defence collaborated with SNECMA in this research programme. The C.450-01 made the first free vertical flight on 6 May 1959 at Melun-Villaroche, but on 25 July, during a transition from vertical to horizontal flight, control of the aircraft was lost and it crashed from 250 ft (75 m). Although the aircraft was destroyed the pilot ejected successfully. Testing the Coléoptère, however, was considered to have been successful despite the accident.

(France)

SUD-EST S.E. 5000 BAROUDEUR

First flight: 1 August 1953
Purpose: Experimental tactical support fighter, designed to operate independently of airfields with long runways
Data: Second prototype
Power plant: One SNECMA Atar 101C turbojet engine (6,173 lb; 2,800 kg st)
Wing span: 32 ft 9½ in (10.00 m)
Length: 44ft 3 in (13.49 m)
Height on trolley: 11 ft 9¾ in (3.60 m)
Wing area: 272 sq ft (25.26 m^2)
Weight loaded: 13,890 lb (6,300 kg)
Max level speed at 19,675 ft (6,000 m): 561 knots (646 mph; 1,040 km/h)
Accommodation: Pilot only
Special design features: Wings had leading-edge sweepback of 38°, and were fitted with leading-edge slats. Trailing-edge flaps inboard of ailerons. Rear section of fuselage detachable for access to engine. Swept fin and rudder, with one-piece tailplane mounted near top of fin. Retractable skids for take-off and landing, with two main skids forward and one under tail. Alternative take-off by independent rocket-powered trolley.

History: The first prototype Baroudeur, powered by an Atar 101B turbojet engine of 5,280 lb (2,394 kg) st, made its maiden flight in August 1953. A second prototype flew for the first time on 12 May 1954, and on 17 July of the same year this aircraft exceeded Mach 1 in a dive. It then demonstrated its ability to operate from uncultivated fields, sand and pebble beaches, and muddy, snow-covered or frozen ground. Three pre-production aircraft, designated S.E.5003s, were then built for service trials. These were fitted with Atar E-4 engines (8,160 lb; 3,700 kg st) and could carry two 1,000 kg bombs under their wings. They attained a maximum speed of 644 knots (742 mph; 1,195 km/h) at S/L and had a service ceiling of 54,950 ft (16.750 m).

(France)

SUD-EST GROGNARD

First flight: 30 April 1950
Purpose: Experimental ground attack aircraft and armament testbed
Photo and data: SE 2410
Power plant: Two Hispano-Suiza Nene 101 turbojet engines (each 4,850 lb; 2,200 kg st)
Wing span: 44 ft 6½in (13.58 m)
Length: 50 ft 7¾ in (15.435 m)
Height: 19 ft 3 in (5.87 m)
Wing area: 496.65 sq ft (46.14 m²)
Weight loaded: 31,967 lb (14,500 kg)
Max level speed at 4,925 ft (1,500 m): 560 knots (645 mph; 1,037 km/h)
Accommodation: Pilot only. Crew of two in SE 2415 version
Special design features: Humped fuselage containing two engines staggered one above the other in the rear fuselage, with the jet exits one above the other in tail. Air inlets in upper fuselage, over wing leading-edge. Sweepback of 47° and 32° on the leading-edges of the wings of the SE 2410 and SE 2415 respectively. Swept tailplane, mounted below jet orifices. Cockpit in extreme nose of SE 2410.
History: Following completion of an 0.582-scale model for testing in the Chalais-Meudon wind-tunnel, the SE 2410 prototype was built. Intended originally as a ground attack aircraft, it was followed by the developed SE 2415 or Grognard II. First flown on 14 February 1951, the Grognard II had a longer forward fuselage and a redesigned cockpit enclosure. Both aircraft were used as armament testbeds, and during their career carried out trials with many weapons, including Matra air-to-air missiles. Several other versions of the design were projected, including the SE 2421 all-weather fighter, which would have been fully equipped for military duty. However, the later projects were abandoned.

(France)

SUD-OUEST ARIEL

First flight (Ariel II): 23 March 1949
Purpose: Experimental helicopter
Photo: SO 1100 Anel I with tailboom
Data: SO 1110 Ariel II
Power plant: Mathis G8 eight-cylinder piston engine (220 hp), driving a Turboméca compressor to provide air for rotor tip-jets
Rotor diameter: 35 ft 5¼ in (10.80 m)
Length of fuselage: 23 ft 5½ in (7.15 m)
Weight loaded: 2,380 lb (1,080 kg)
Max level speed: 92 knots (106 mph; 170 km/h)
Hovering ceiling: 2,950 ft (900 m)
Duration: 1½ hours
Accommodation: Two seats
Special design features: The stubby Ariel I had crew compartment in nose with engine area aft and twin fins at rear. Ariel II had long tailboom supporting reshaped twin vertical tail surfaces. Ariel III had alternative Turboméca Artouste or Arius turbine engines (220 or 275 shp), a single fin and rudder and side-mounted air intakes. Rotor drive was provided in Ariel helicopters by feeding compressed air from the power plant through the rotor hub and the hollow rotor blades to small combustion chambers at the blade tips. The air was fed at low pressure through the hub but was centrifuged in its passage through the blades to the combustion chambers, where fuel was added and ignited electrically.
History: The first helicopter of the series was the SO 1100 Ariel I, which was tested initially in 1947. Although, for these tests, the vertical tail surfaces were fitted immediately aft of the engine area, the Ariel I was later fitted with a long tailboom and a single rudder. It was followed by the SO 1110 Ariel II, which retained a generally similar configuration to the developed Ariel I but had the underfuselage strake removed and was fitted with new twin vertical tail surfaces. The SO 1120 Ariel III, which first flew on 18 April 1951, was powered by a turbine engine and had large side-mounted air intakes. A single fin and rudder was fitted to the tailboom and, because of the weight saved by the installation of the turbine-compressor unit, the Ariel III was able to have three seats.

SUD-OUEST SO 1310 FARFADET

(France)

First flight (as helicopter): 8 May 1953
Purpose: Experimental convertiplane
Power plant: One Turboméca Arius II turbo-compressor (360 hp), which supplied compressed air to the jet-driven rotor. One Turboméca Artouste II turboprop engine (360 hp), which drove a variable-pitch propeller in the nose of the fuselage
Rotor diameter: 36 ft 9 in (11.20 m)
Wing span: 20 ft 8 in (6.30 m)
Cruising speed: about 130 knots (150 mph; 240 km/h)
Accommodation: Crew of two seated side by side with dual controls. Either three passengers, two stretchers or freight
Special design features: Gyrodyne-type aircraft, with conventional aeroplane fuselage. Power provided by two independent turbine units. All-metal three-blade jet-driven rotor, similar to that of the SO 1120 Ariel III, with small combustion chambers at the blade tips. Narrow-chord wings (used as fuel tanks) and non-retractable tricycle landing gear.
History: The Farfadet was the first French aircraft of the Gyrodyne type. This configuration enabled it to take-off vertically, hover and land vertically in the manner of a helicopter, with powered rotor, and also to fly at a speed greater than that of a normal helicopter, because of its fixed wings and a propeller which was independent of the rotor. In this aircraft, translation from rotary-wing to fixed-wing flight was accomplished without any change in the exterior configuration. During forward flight the rotor continued to turn in autorotation, providing some lift, while the fixed wings provided the primary lift. The Farfadet achieved its first transitional flight on 1 July 1953.

(France)

SUD-OUEST SO 6000 TRITON

First flight: 11 November 1946
Purpose: Experimental trainer
Photo: SO 6000-03
Data: SO 6000-04
Power plant: One Hispano-Suiza Nene 101 turbojet engine (4,850 lb; 2,200 kg st)
Wing span: 32 ft 8 in (9.96 m)
Length: 34 ft 2 in (10.41 m)
Wing area: 161.46 sq ft (15.0 m²)
Weight loaded: 10,053 lb (4,560 kg)
Max level speed at S/L: 515 knots (593 mph; 955 km/h)
Service ceiling: 39,375 ft (12,000 m)
Accommodation: Two persons side by side, with dual controls

Special design features: Bulky oval-section fuselage. Thin short-span wings, tapering in planform and with rounded tips. First aircraft had air intake at nose. The first Nene-powered aircraft (SO 6000-04) had extra fuselage side-mounted intakes behind the cockpit. The following three aircraft had only the side-mounted intakes.

History: The SO 6000 was the first French jet-propelled aircraft to fly. Development began secretly in 1943 and the first prototype was built in 1945. Five prototypes were ordered, four originally to be fitted with Derwent engines. The first, known as the SO 6000-01, was fitted with a German Junkers Jumo 004 B2, the only satisfactory type of turbojet engine then available in France; but the later aircraft were all powered with Nene engines. The first Nene-powered aircraft, the SO 6000-04, made its first flight on 19 March 1948.

(France)

SUD-OUEST SO 9000 TRIDENT

First flight: 2 March 1953
Purpose: Mixed-power research aircraft
Photo: Original version with Marboré engines
Power plant: Powered initially by two Turboméca Marboré II turbojet engines (each 880 lb; 400 kg st), mounted at the wingtips; these were replaced by two Dassault M.D. 30 Viper ASV.5 turbojet engines (each 1,640 lb; 744 kg st). Later, an SEPR.481 rocket motor (total thrust 9,920 lb; 4,500 kg) was installed in rear fuselage
Wing span: 26 ft 8¾ in (8.15 m)
Length: 45 ft 11¼ in (14.00 m)
Wing area: 99.03 sq ft (9.20 m²)
Weight loaded: 12,125 lb (5,500 kg)
Max level speed at 36,000 ft (11,000 m): 917 knots (1,056 mph; 1,700 km/h)
Service ceiling: about 59,000 ft (17,985 m)
Ferry range (turbojets only): 318 nm (366 miles; 590 km)
Accommodation: Pilot only
Special design features: Shoulder-mounted short-span wings of constant chord and thin section. Ailerons locked after first flight trials had proved adequacy of using tail surfaces for all control. Tail surfaces made up of three one-piece all-moving planes. Vertical surface acted as rudder; the other two had marked anhedral and provided both roll and pitch control. Circular-section fuselage with three-barrel rocket motor at rear. Tail braking parachute. Entire nose of fuselage jettisonable in an emergency.
History: Initial design began in 1948, and the development contract for the first prototype was signed in 1951. First-stage testing of the Trident was with Turboméca wingtip engines; these were later replaced by Dassault engines. Subsequently, an SEPR rocket motor was installed in the aircraft. Powered by the Turboméca turbojets only, the Trident made its maiden flight in March 1953. Its first flight with rocket power took place on 4 September 1954. The Trident resumed flying with the higher-powered turbojets on 18 March 1955, and soon afterwards exceeded Mach 1 by a large margin in a shallow dive, without bringing the rocket motor into use. Tests were then begun under rocket power, and on 3 April 1955 the Trident exceeded the speed of sound during the last stage of its climb, using only a fraction of the available thrust. Eventually, the Trident exceeded Mach 1.5 in both level and climbing flight. A second aircraft was built, but this crashed on its maiden flight in September 1953. The SO 9050 Trident II prototype interceptor was a direct development of the SO 9000 Trident.

(France)

SUD-OUEST ESPADON (SWORDFISH)

First flight: 12 November 1948
Purpose: Experimental interceptor and research aircraft
Photo: SO 6025
Data: SO 6021
Power plant: One Hispano-Suiza-built (Rolls-Royce) Nene turbojet engine (5,000 lb; 2,270 kg st)
Wing span: 34 ft 9¼ in (10.60 m)
Length: 49 ft 3 in (15.00 m)
Wing area: 292.78 sq ft (27.2 m²)
Weight loaded: 15,430 lb (7,000 kg)
Max level speed at S/L: 539 knots (621 mph; 1,000 km/h)
Service ceiling: 42,650 ft (13,000 m)
Accommodation: Pilot only
Special design features: Sweptback wings with rounded tips. High cockpit canopy. Various power plant installations, including rocket motor (see below). Ejection seat.
History: The design of the SO 6020 Espadon began in 1945, as a single-seat jet interceptor to meet requirements laid down in the first French post-war official military aircraft programme. From the original prototype several experimental versions were developed, the last being the SO 6026 which flew for the first time in 1951. The various versions offered as follows:

6020-01. First prototype. Jet interceptor fighter design. One Nene engine with ventral air intake. Wing area 271 sq ft; gross weight 17,637 lb. Subsequently modified to -02 standard and fitted, in 1952, with two wingtip-mounted Turbomeca, Marbore turbojet engines, for research with this configuration.

6020-02. Second prototype. Flush side air intakes. Raised fin with increased area. First flown 30 December 1949. Later converted to 6026.

6020-03. Third prototype. Modified and redesignated 6025 before completion.

6021. Development of 6020-02. Lighter structure. Smaller pressurised and air-conditioned cockpit, and ejection seat. Increased wing area. Leduc-Jacottet hydraulic servo controls. Armed with six 20 mm or four 30 mm cannon. Said to have exceeded Mach 0.95 in a dive from 39,375 ft. First flown 3 September 1950.

6025. This was the redesignated 6020-03. Ventral air intake with double entries. SEPR 251 bi-propellant rocket, with jettisonable rocket fuel tanks in common fairing between air intakes beneath fuselage. Same equipment and wing area as 6021. Provision in fuselage nose for several types of cameras. First flown 28 December 1949

6026. This was the 6020-02 modified and redesignated. Rocket motor in rear of fuselage under jet tailpipe. Part of internal tankage and added wingtip tanks held rocket propellants. First flown 15 October 1951.

(France)

SUD-OUEST SO M.2

First flight: 13 April 1949
Purpose: Flying "scale model" of the SO 4000 experimental bomber which was used for high-speed research
Power plant: One Rolls-Royce Derwent 5 turbojet engine (3,500 lb; 1,587 kg st)
Wing span: 31 ft 2 in (9.50 m)
Length: 32 ft 5¾ in (9.90 m)
Wing area: 193.75 sq ft (18.0 m²)
Weight loaded: 10,363-11,905 lb (4,700-5,400 kg)
Max level speed: over 539 knots (621 mph; 1,000 km/h)
Accommodation: Pilot only
Special design features: Laminar-flow wings with very thick flush-riveted skin, swept at 31° at the main spar. Flaps extended over most of the trailing-edge of the wings. Small ailerons at the tips were interconnected with spoilers to provide lateral control. Leading-edge slots fitted. Air intake ducts on each side of the fuselage. Landing gear made up of a nosewheel, three wheels in tandem beneath the fuselage and small wheels under the wingtips.
History: Following experiments with the M.1, a half-scale model of the SO 4000 bomber which was flown as a glider, the M.2 appeared as a powered version of the M.1. Both were used to gain experience with a number of design features which, up to that time, had not been widely experimented with in France. The M.1 had first flown in 1948. In May 1950 the M.2 became the first French aircraft to exceed 1,000 km/h in level flight. It entered the second phase of its tests on 15 September 1951, following modifications which included the fitting of hydraulic servo controls and wingtip tanks, and provision for the installation of solid-propellant rockets for thrust augmentation. The test instruments carried by the M.2 provided considerable information on the effectiveness of various control systems at high subsonic speeds at high altitude.

(UK)

SUPERMARINE TYPE 510

First flight: 29 December 1948
Purpose: To investigate flight characteristics at high subsonic speeds
Power plant: One Rolls-Royce Nene turbojet engine (5,000 lb; 2,270 kg st)
Wing span: 31 ft 8½ in (9.67 m)
Length: 38 ft 1 in (11.61 m)
Height (tail down): 8 ft 9¾ in (2.69 m)
Wing area: 273 sq ft (25.36 m²)
Max level speed: about 564 knots (650 mph; 1,046 km/h)
Accommodation: Pilot only
Special design features: Basically a Supermarine Attacker fitted with sweptback wings and tailplane. Supermarine laminar-flow wings, swept at 40° at quarter chord.
History: Designed to Specification E.41/46, the Type 510 was sent, after its initial flights, to the A and AEE at Boscombe Down in October 1949 for handling trials. Unfortunately, the aircraft was found to vibrate at low engine speeds and was returned to its makers in November. Handling trials were carried out during the first three months of 1950, after modifications had been made to give the aircraft an Attacker-type nose, split trailing-edge flaps, underwing dive-recovery flaps, etc. During November 1950, the Type 510 successfully completed deck-landing trials on board HMS **Illustrious.** For these trials the aircraft was fitted with an arrester hook and RATOG. (This was the first landing of a sweptwing aircraft on an aircraft carrier). A second 510 was built, with an afterburner in a lengthened rear fuselage. This aircraft was designated Type 535, and became a predecessor of the Supermarine Swift fighter, after it had been given an extended nose and other modifications, including a tricycle landing gear (a retractable twin tailwheel unit was also fitted to make a tail-down landing possible, with a corresponding increase in the angle of attack of the wings for braking effect).

(USSR)

SUKHOI Su-15 "FLAGON B"

First flight: Unknown
Purpose: Research and development prototype STOL fighter
Photo: Flagon B with lift-jet doors open
Power plant: Two afterburning turbojet engines, with variable-area nozzles, mounted side by side in rear fuselage. Three lift-jet engines mounted vertically in the centre-fuselage
Wing span (estimated): 35 ft 0 in (10.67 m)
Length (estimated for Su-15 Flagon A): 68 ft 0 in (20.5 m)
Max level speed above 36,000 ft (11,000 m) clean, (Su-15 Flagon A, estimated): Mach 2.5
Accommodation: Pilot only
Special design features and History: The Sukhoi prototype STOL fighter demonstrated at Domodedovo on 9 July 1967 differed little from the standard first-line Su-15 "Flagon A" type. The main difference was the installation of three lift-jet engines, mounted vertically in tandem under two rearward-hinged intake doors in the centre-fuselage, between the air intake trunks. There were longitudinal slots in the doors and transverse louvres in the panels under the fuselage, beneath the lift-jets. The usual dielectric nosecone appeared to be replaced by a fairing, painted black, with the paint area extending further aft than the normal cut-off line of the dielectric cone. Black triangles forward of the engine air intakes were painted on, perhaps to give the impression of half-cone centrebodies. The wing span was greater than that of the "Flagon A", giving compound sweep at the tips, from a station just outboard of the boundary-layer fence each side. The wing root leading-edge was swept forward, giving increased chord immediately adjacent to each air intake trunk. There was a large scoop on the rear fuselage above each jet nozzle, probably for additional afterburner cooling.

(Germany)

VFW-FOKKER H3

First flight: Spring of 1970
Purpose: Experimental rotorcraft
Power plant: One Allison 250-C18 turboshaft engine (317 shp)
Rotor diameter: 28 ft 6½ in (8.70 m)
Length of fuselage: 24 ft 2¼ in (7.37 m)
Max payload: 595 lb (270 kg)
Max T-O weight: 2,134 lb (968 kg)
Max cruising speed at S/L (estimated): 135 knots (155 mph; 250 km/h)
Service ceiling (estimated): 13,120 ft (4,000 m)
Max endurance (estimated): 3 hrs as autogyro, 2 hrs as helicopter
Accommodation: One pilot and seating for two passengers
Special design features and History: The H3 was to have been the first in a family of helicopters with a "cold" tip-jet driven rotor developed by VFW-Fokker, embodying experience gained from the experimental WFG-H2 and utilising the same basic principles.

A turbo-compressor provided compressed air to tip-drive the three-blade rotor, without blade-tip combustion. This eliminated the need for conventional transmission and drive-shaft systems, hydraulic systems, clutches and torque compensation, while allowing full rotor autorotation to be maintained in the event of engine failure. The method of propulsion originally proposed proved unsuitable for rotorcraft of this size; thus, the flight tests were concentrated on system development.

Two prototypes of the VFW-H3 were completed in 1968, and underwent extensive ground testing. Flight testing of the H3-E1 began in 1970, and flight testing of a second H3 prototype, which had a more powerful engine and an improved compressor, began in early 1972. The result of ground and flight testing, and further theoretical investigations, led to design of the improved H4, but the whole programme was later suspended.

(Germany)

VFW-FOKKER VAK 191B

First flight: 10 September 1971
Purpose: V/STOL experimental strike and reconnaissance fighter
Power plant: Two Rolls-Royce RB.162-81 lift-jets (each 5,577 lb; 2,530 kg st) mounted vertically in the fuselage, and one Rolls-Royce/MTU RB.193-12 vectored-thrust turbojet (approx 10,150 lb; 4,603 kg st) for forward propulsion
Wing span: 20 ft 2½ in (6.16 m)
Length overall: 53 ft 7 in (16.335 m)
Height: 14 ft 1 in (4.295 m)
Wing area: 134.5 sq ft (12.50 m²)
Max T-O weight: 19,840 lb (9,000 kg)
Accommodation: Pilot only
Special design features: Short-span wings with 12° 30′ anhedral from roots. Incidence 1° 30′. Sweepback at quarter-chord approx 40°. Multi-spar fail-safe structure of aluminium alloy, including ailerons and trailing-edge flaps. Forward exhaust doors of both lift engines extend to act as airbrakes. Sweptback tail unit with one-piece all-moving tailplane. Tandem-type main landing gear, with outrigger wheels at wingtips. Stabilisation of aircraft by "puffer-jets" at nose, tail and wingtips.
History: To meet the German requirement VAK 191B for a subsonic VTOL tactical reconnaissance fighter to replace the Fiat G91, the former Focke-Wulf company produced a design study under the designation FW 1262. A development of this project was initiated jointly by VFW and Fiat of Italy, under an agreement signed by the German and Italian Defence Ministers in early 1964. In 1968 the Italian government withdrew from the programme, which was then continued by VFW-Fokker; Aeritalia remained associate subcontractor responsible for the previously-agreed portions of the airframe.

The number of prototypes to be completed was later reduced from six to three. The first was rolled out in April 1970 and made its initial flight after a period of tethered hovering trials. During 1972 the three VAK 191B prototypes continued the flight test programme to explore the complete performance spectrum, and on 26 October 1972 the first vertical-to-horizontal transition was made successfully. During the test the aircraft flew at a speed of 300 knots (345 mph; 556 km/h) and the jet-lift engines were shut down and restarted in flight.

(UK)

VICKERS TYPE 618 NENE-VIKING

First flight: 6 April 1948
Purpose: Experimental jet-propelled conversion of a Vickers Viking 1B commercial aircraft, to test new engines and to produce data for high-altitude flying by airliners
Power plant: Two Rolls-Royce Nene 1 turbojet engines (each 5,000 lb; 2,268 kg st)
Wing span: 89 ft 3 in (27.20 m)
Length (tail up): 65 ft 2 in (19.86 m)
Height (tail up): 24 ft 0 in (7.32 m)
Weight empty: 21,050 lb (9,548 kg)
Weight loaded: 33,500 lb (15,195 kg)
Max level speed at 30,500 lb (13,834 kg) mean weight, at 10,000 ft (3,050 m): 406 knots (468 mph; 753 km/h)
Service ceiling, above weight: 44,000 ft (13,400 m)
Max still air range at 10,000 ft (3,050 m), above weight at 341 knots (393 mph; 632 km/h): 300 nm (345 miles; 555 km)
Accommodation: Provision for crew of 4 and 24 passengers, plus baggage, etc
Special design features: Generally similar to standard Vickers Viking 1B, except for turbojet engines in long underslung nacelles and heavier metal skin on the wings and tail surfaces.
History: This was the first completely jet-driven commercial aeroplane to fly. It took off for the first time from Wisley, Surrey, and on 25 July 1948 completed a London-to-Paris flight in 34 mins 7 sec. After a period of tests the type 618 was converted back to its original form and was used as a freighter.

(USA)

WILLIAMS RESEARCH WASP

First flight: 1974?
Purpose: Experimental flying platform
Photo: Wasp attached to safety tether line
Power plant: One Williams WR19-9 miniature turbofan engine (700 lb; 317 kg st)
Empty weight: 270 lb (123 kg)
Max level speed: up to 52 knots (60 mph; 96 km/h)
Accommodation: Two persons, standing
Special design features: Essentially a platform with the engine mounted vertically.
History: On 26 January 1970, the Bell Aerospace Company Division of Textron Inc announced that it had granted to Williams Research Corporation a licence to manufacture, use and sell certain small lift device systems in the US and Canada. They included the Jet Flying Belt which Bell had developed for the US Army. Since that time Williams has been working on a new and more advanced version of the Bell concept, and announced on 14 February 1974 that it had tested successfully a two-man turbine-powered flying platform. Known as the WASP (Williams Aerial Systems Platform), its flight tests were conducted with the vehicle attached to a safety tether line, under a US Navy contract to demonstrate its suitability to meet a US Marine Corps STAMP (Small Tactical Aerial Mobility Platform) requirement. In operation, the pilot merely has to mount the platform, start the engine and fly away, using simple hand controls so designed that the vehicle can be controlled with one hand in flight. It is anticipated that the WASP will prove capable of carrying two men at speed for a duration of approximately 30 minutes. The vehicle will be able to accelerate rapidly, move in any direction, hover and rotate on its own axis. The company believes that WASP will be suitable for a number of military and civil applications, including law enforcement, firefighting, rescue and medical aid.

(USSR)

YAKOVLEV Yak-36

First flight: Unknown
Purpose: VTOL research aircraft
Power plant: Two turbojet engines, mounted side by side at bottom of front fuselage and each exhausting through a large-diameter louvred and gridded vectored-thrust nozzle. Bleed air supply to "puffer-pipe" reaction control nozzles located at the tail, at end of a massive nose-probe and in each wingtip fairing, for control in hovering and low-speed flight
Wing span, between centre-lines of wingtip fairings (estimated): 27 ft 0 in (8.25 m)
Length overall (estimated): 57 ft 6 in (17.50 m)
Length of fuselage excl nose-probe (estimated): 41 ft 0 in (12.50 m)
Height overall (estimated): 14 ft 9 in (4.50 m)
Max level speed: Subsonic
Accommodation: Pilot only
Special design features: Cropped-delta wings, with anhedral from roots. Sweepback approx 40° on leading-edges. Entire trailing-edge hinged, as flaps and ailerons. Wide fuselage of elliptical cross-section, with divided ram air intake in nose. Two large blister fairings on front fuselage, under each engine. Full-width backward-hinged door under nose to reduce possibility of air recirculation into intakes during take-off and landing. Smaller rearward-hinged door under centre-fuselage, forward of nozzles, prevents undesirable interaction of exhaust gases under fuselage, in conjunction with two longitudinal strakes forward and inboard of nozzles. Double-hinged panel forward of main landing gear unit protects it from hot exhaust gases and is also used as an airbrake. Swept tail unit, with fixed-incidence tailplane. Retractable tandem-type main landing gear, with two small balancer wheels which retract into wingtip fairings.
History: This aircraft is much less refined than the British Hawker Siddeley Harrier and must be regarded as a purely experimental design. Only one of two Yak-36s sent to the air display at Domodedovo in July 1967 took part in the flying programme. Code-named "Freehand" by NATO, the aircraft took off vertically, performed a transition at a height of about 160 ft (50 m), made a circuit of the airfield, including a high-speed fly-past, and ended with a 180° hovering turn before making a vertical landing. Although about a dozen Yak-36s are believed to have been built, the type is unlikely to have been developed beyond the research and development phases. However, one Yak-36 is reported to have carried out sea trials from a specially-installed pad on the helicopter carrier *Moskva;* and it is known that the Soviet Navy required a V/STOL combat aircraft for its new *Kuril* class carriers. It is thought that Yakovlev has evolved such an aircraft from the Yak-36.

(Yugoslavia)

YUGOSLAV GOVERNMENT FACTORIES TYPE 451M

First flight: Late 1952
Purpose: General research into jet-propelled aircraft and prototype for projected military designs
Power plant: Two Turboméca Palas turbojet engines (each 330 lb; 150 kg st)
Wing span: 21 ft 11¾ in (6.7 m)
Length: 24 ft 3½ in (7.4 m)
Height: 7 ft 7½ in (2.32 m)
Weight loaded: 2,976 lb (1,350 kg)
Max level speed: 254 knots (292 mph; 470 km/h)
Service ceiling: 27,875 ft (8,500 m)
Range: over 162 nm (186 miles; 300 km)
Accommodation: Pilot only
Special design features: Jet-powered conversion of the Type 451 light twin-engined research aircraft, which was powered by two Walter Minor 6-III piston engines (each 160 hp). Fuselage had nigh-mounted cockpit canopy and an undernose fairing for a 20 mm cannon. Six rockets could also be carried. Engines mounted under the straight wings.
Inward-retracting main landing gear and semi-retracting tailwheel.
History: Developed from the Type 451 piston-engined aircraft, which featured a prone position for its pilot, the first Type 451M was built at the Ikarus factory at Zemun, becoming the first jet-propelled aircraft of Yugoslav design. Its designer, Major Dragoljub Beslin, continued the development of the basic type and several versions were built for testing. These included the J-451MM Strsljen (Hornet) single-seat close support aircraft, the S-451MM Matica (Queen Bee) tandem two-seat basic trainer, and the T-451MM Strsljen II single-seat trainer. All were generally similar to each other, except for their forward fuselages, cockpits and minor tail differences. The 451MMs also differed from the original Type 451M jet in that the engines were Turboméca Marboré IIs, mounted inside and not underneath the wings.

(Yugoslavia) YUGOSLAV GOVERNMENT FACTORIES TYPE 452-2

First Flight: 24 July 1953
Purpose: Experimental aircraft
Power plant: Two Turboméca Palas turbojet engines (each 353 lb; 160 kg st)
Wing span: 17 ft 2¾ in (5.25 m)
Length: 19 ft 7 in (5.97 m)
Height: 5 ft 9¾ in (1.77 m)
Weight loaded: over 2,335 lb (1,059 kg)
Max level speed: 421 knots (485 mph; 780 km/h)
Min speed (estimated): 102 knots (118 mph; 190 km/h)
Duration: 75 min
Accommodation: Pilot only

Special design features: Swept wings. Vertical tail surfaces mounted at ends of twin booms, with the Vee tailplane between the fins and supported at its centre by a dorsal fin on aft end of fuselage. Turbojet engines mounted one above the other at rear end of fuselage. Separate air inlets provided for each engine, in the wing roots for lower engine and on the sides of the rear fuselage for upper engine. Tricycle-type retractable landing gear. An experimental light alloy was used for the whole structure. Two 12.7 mm machine-guns.

History: The 452, designed by Major D. Beslin, was the second type of experimental light jet aeroplane to be built in Yugoslavia (the first being the 451M). Two prototypes were built at the Ikarus factory to provide data for a projected light attack aircraft.

INDEX

A

Aérocentre N.C.3020 Belphégor	7
Air Horse	55
Ambrosini Sagittario	9
Ariel	227
Armstrong Whitworth A.W.52	11
Arsenal VG 70	13
Avro Ashton	15
Avro Type 707	17

B

B-35	177
BAC:	
221	19
TSR 2	21
(Hunting) H.126	23
Baroudeur	223
Bell:	
X-1	25
X-2	27
X-5	29
X-14	31
X-22A	33
XV-3	35
XV-15	37
Air Cushion Landing System	71
Belphégor	7
Boeing Vertol:	
Model 347	39
VZ-2A	41
Bolkow Bo 46	43
Boulton Paul P.111 and P.120	45
"Bounder"	157
Bratukhin Omega	47
Bristol T.188	49

C

Chance Vought XF5U-1	51
Cierva:	
W.9	53
Air Horse	55
Coléoptère	221
Convair:	
Sea Dart	57
XB-46	59
XF-92A	61
XFY-1	63
Curtiss-Wright:	
X-100	65
X-19A	65

D

Dassault Balzac:	
V-001	67
III-V	67
De Havilland DH 108	69
De Havilland Canada XC-8A	71
Doak VZ-4DA	73
Dornier Do 31E	75
Douglas:	
D-558-7 Skystreak	77
D-558-8 Skyrocket	79
X-3	81
XB-42	83
XB-43	83
Draken	205

E

E-166	145
Espadon	235
EWR VJ 101 C	85

F

F-107A	171
Fairchild:	
VZ-5FA	87
XC-120 Pack-Plane	89
Fairey:	
F.D.1	91
Delta 2	93
Gyrodyne	95
Rotodyne	97
V.T.O. Project	99
"Faithless"	147
Farfadet	229
"Fishbed-G"	151
"Flagon-B"	239
Fouga C.M. 88-R Gemeaux	101

G

Gemeaux	101
Gerfaut	165
Griffon	167
Grognard	225
Gyrodyne	25

H

Handley Page:	
H.P. 88	105
H.P. 115	107
H.P. 75 Manx	103
Hawker P.1052	109
Hawker Siddeley P.1127 Kestrel	111
Hiller X-18	113
HL-10	183
Hughes:	
XH-17	115
XV-9A	117
Hunting H.126	23

I

I.Ae.27 Pulqui	119
I.A. 38	121

K

Kamov Ka-22 Vintokryl	123
Kestrel	111

L

Leduc:	
0.10	125
0.21	127
0.22	127
Lockheed:	
XFV-1	129
XH-51A	131
XV-5A Hummingbird	133

M

M2-F2	185
M-52	153
M-52 "Bounder"	157
Manx	103
Martin Marietta:	
X-24A	135
X-24B	137
McDonnell:	
XF-85 Goblin	139
XF-88B	141
XV-1	143
Mikoyan:	
E-166	145
"Faithless"	147
MiG-21 Analogue	149
MiG-21 "Fishbed-G"	151
Miles M-52	153
Mississippi State University	
Marvel and Marvelette	155
Myasishchev M-52 "Bounder"	157

N

NASA/DITC Augmenter Wing Jet Stol	159
NASA Supercritical Wing	161
Nene- Viking	247
Nord:	
500	163
Gerfaut	165
1500 Griffon	167
1601	169
North American:	
F-107A	171
X-15A	173
XB-70A Valkyrie	175
Northrop:	
B-35	177
YB-49	177
X-4 Bantam	179
XP-79B	181
Northrop/NASA:	
HL-10	183
M2-F2	185

O

Omega	47

P

P.111	45
P.120	45
P.1052	109
P.1127 Kestrel	111
Payen PA 49	187
Piasecki:	
16H-1A Pathfinder II	189
VZ-BP (B) Airgeep II	191
Pulqui	119

R

Republic XF-91	193
RFP (Lippisch) X-113 Am Aerofoil Boat	195
Rotodyne	97
Ryan:	
VZ-3RY Vertiplane	197
X-13 Vertijet	199
XF-2R-1	201
XV-5A and Xv-5B	203

S

Saab 210 Draken	205
Sagittario	9
Saunders-Roe:	
SR.A/1	207
S-R. 53	209
Sea Dart	57
Short:	
S.B. 4 Sherpa	211
S.B.5	213
S.C.1	215
Sikorsky:	
S-72	217
XH-59A	219
Skystreak	77
Skyrocket	79
SNECMA C.450-01 Coléoptère	221
Sud-Est:	
S.E. 5000 Baroudeur	223
Grognard	225
Sud-Ouest:	
Ariel	227
SO 1310 Farfadet	229
SO 6000 Triton	231
SO 9000 Trident	233
Espadon	235
SO M.2	237

Sukhoi Su-15 "Flagon-B"	239	X-5	29	YB-49	177
Supermarine Type 510	241	X-13 Vertijet	199	Yugoslav Government Factories:	
		X-14	31	Type 451M	253
		X-15A	173	Type 452-2	255

T

T.188	49
Trident	233
Triton	231
TSR 2	21

X-18	113
X-22A	33
X-24A	135
X-24B	137
X-100	65
X-194	65
XB-42	83
XB-43	83
XB-46	59
XB-70A Valkyrie	175
XC-8A	71
XC-120 Pack-Plane	89
XF2R-1	201
XF5U-1	51
XF-85 Goblin	139
XF-88B	141
XF-91	193
XF-92A	61
XFV-1	129
XFY-1	63
XH-17	115
XH-51A	131
XH-59A	219
XP-79B	181
XV-1	143
XV-3	35
XV-5A	203
XV-5A Hummingbird	133
XV-5B	203
XV-9A	117
XV-15	37

V

V-001	67
Vak 191 B	245
VFW-Fokker:	
H3	243
Vak 191B	245
VG 70	13
Vickers Type 618 Nene-Viking	247
Vintokryl	123
VJ 101C	85
VZ-2A	41
VZ-3RY Vertiplane	197
VZ-4DA	73
VZ-5FA	87
VZ-BP (B) Airgeep II	191

W

W.9	53
Williams Research Wasp	249

X

X-1	25
X-2	27
X-3	81
X-4 Bantam	179

Y

Yakovlev Yak-36	251

PHOTO CREDIT LIST

Flight: 44, 52
S.P. Peltz: 18
Ministry of Supply: 54, 72, 246
Howard Levy: 86
Charles E. Brown: 94
Reportage Airmondial: 131, 100
B.M. Service: 104
Keystone press agency: 106
Associated Press: 114, 156, 226
Novosti Press Agency: 122, 148
NASA: 134, 158, 160, 172
New York Times: 138
Repülés: 146
Tass: 150, 250
Geoffrey Goddard: 214
Butler-Green Aviation Photo Service: 126
Air Pictorial: 152
Pilot Press: 6, 8, 46, 168.